NEW PATHWAYS IN SCIENCE

NEW PATHWAYS IN SCIENCE

by

SIR ARTHUR EDDINGTON

UNITY SCHOOL LIBRARY
Unity Village
Lee's Summit, Missouri 64063

ANN ARBOR PAPERBACKS
THE UNIVERSITY OF MICHIGAN PRESS

First edition as an Ann Arbor Paperback 1959

All rights reserved

First published by the Cambridge University Press

Reprinted by special arrangement

Manufactured in the United States of America

CONTENTS

Preface		page vii
Chapter I	Science and Experience	1
II	Dramatis Personae	27
III	The End of the World	50
IV	The Decline of Determinism	72
V	Indeterminacy and Quantum Theory	92
VI	Probability	110
VII	The Constitution of the Stars	135
VIII	Subatomic Energy	160
IX	Cosmic Clouds and Nebulae	184
X	The Expanding Universe	206
XI	The Constants of Nature	229
XII	The Theory of Groups	255
XIII	Criticisms and Controversies	278
XIV	Epilogue	309
Index		327

PLATES

Plate 1	Electrons and Positrons	Facing page 28
	By permission of Prof. P. M. S. Blackett	
2	Gaseous Nebula (Cygnus)	184
3	Dark Nebulosity—The Horse's Head	202
4	Spiral Nebula (Canes Venatici)	206

PREFACE

THIS volume contains the Messenger Lectures which I delivered at Cornell University in April and May 1934. Chapters II and VIII have been added; the remaining chapters correspond to the twelve lectures of the course. It was one of the conditions of the lectureship that the lectures should be published.

Except for a small book on the Expanding Universe, my last spell of writing was about six years ago, when *Stars and Atoms* (1927), *The Nature of the Physical World* (1928) and *Science and the Unseen World* (1929) practically exhausted all that it was then in my mind to say. A scientific writer is placed in a difficulty by his earlier books; either his new book will appear as a rather disjointed addendum to them, or he must perfunctorily go over again a great deal of matter which he has no wish to rewrite. Being unwilling to adopt the second alternative, I determined to make what I could of whatever had come to my mind in the last six years. Accordingly I spoke at Cornell on a variety of topics, using as a nucleus the material contained in a number of addresses and lectures which I had had occasion to deliver since 1929, and adding other subjects to which I had been giving attention. The general plan was that each lecture should have a separate theme, except that Indeterminism was spread over two lectures. The choice of subjects has allowed a certain amount of continuity of treatment; but there has been no attempt to provide a systematic introduction to modern scientific thought. Perhaps the biggest gap is the absence of any account of the elementary ideas of the theory of relativity;

I could not bring myself to go over again the ground covered in Chapters I, II, III, VI, VII of *The Nature of the Physical World* altering the treatment and illustrations merely for the sake of alteration.

In the opening lecture I try to explain the philosophical outlook of modern science, as I understand it, and show how the scientific picture of the world described in physics is related to the "familiar story" in our minds. Chapter II is an interpolation containing a summary of our knowledge of atomic physics, etc., which some readers may find necessary for an understanding of subsequent chapters and others may find useful as a reminder. Then follow four lectures which have something in common; they are concerned with the consequences of the statistical type of law, first introduced into physics in the subject of thermodynamics, which has in recent years completely driven out the older causal type of law from the foundations of physics. The last of these four lectures, on Probability, has besides its application to statistical law a more elementary interest.

Then follows a complete change of subject, and the next four lectures are devoted to astrophysics. Starting with the sun and familiar stars, we advance to greater distances till we reach the system of milliards of galaxies which constitutes the universe. This last subject has been treated more fully in my recent book *The Expanding Universe*; I here give a much shorter account. In this lecture (Chapter X) we meet the elusive "cosmical constant" which takes us back to the fundamental conceptions of physics again for the next two chapters. Chapter XI is, I realise, much too severe for this kind of book; I can only plead that the subject which has occupied me for the last five years, almost to the exclusion of any other research, was bound to spill over into any course of lectures I might give. The next lecture, on Theory of Groups, was something of an experiment; but it, more nearly than any other part of the book, touches the key-note of scientific philosophy.

The chapter "Criticisms and Controversies" may by its title lead the reader to expect a comprehensive series of answers to the multitudinous points raised by critics and reviewers, and by many who have contributed valuable discussion of the views which I have advocated. I think that a little reflection will show that this was impracticable with any reasonable allotment of space. If a criticism can be answered briefly and decisively it seems scarcely worth while to inform the world in general that so-and-so has raised it. If it is more arguable, a lengthy explanation and discussion of it is usually necessary. For the most part I am content to think that if my contentions are of value they will find their proper level without continual parental intervention to save them from determined opponents—and sometimes from over-enthusiastic friends. But I would express here my gratitude for many articles by philosophers and others courteously discussing my writings. Sometimes I have appreciated the justice of the criticism, and it has had its due influence in maturing my views. Often I would have liked to write a reply in the hope of advancing an understanding on both sides; but such a reply requires at least as much time and care as an independent article, and with rare exceptions I have had to let the opportunity go by. In the concluding lecture I return again to the philosophical outlook of Chapter I, but this time I refer to that part of "the problem of experience" which the methods of physics do not profess to treat. Parts of this lecture are taken from an address which I gave in a broadcast symposium on Science and Religion.

As usual, notwithstanding my efforts to simplify things, I have to impose a rather heavy strain on the attention of the reader. Since the chapters are to a considerable extent independent, the difficulty tends to increase towards the ends of the chapters. There is hope of a respite when the next chapter begins.

These lectures carry for me happy memories of the weeks which I spent in Cornell University. To the friends who welcomed me, and to the large audiences who encouraged me, I dedicate them gratefully.

A. S. E.

CAMBRIDGE
September 1934

CHAPTER I

SCIENCE AND EXPERIENCE

Does the harmony which human intelligence thinks it discovers in Nature exist apart from such intelligence? Assuredly no. A reality completely independent of the spirit that conceives it, sees it or feels it, is an impossibility. A world so external as that, even if it existed, would be for ever inaccessible to us. What we call "objective reality" is, strictly speaking, that which is common to several thinking beings and might be common to all; this common part, we shall see, can only be the harmony expressed by mathematical laws.

POINCARÉ, *The Value of Science*.

I

As a conscious being I am involved in a story. The perceiving part of my mind tells me a story of a world around me. The story tells of familiar objects. It tells of colours, sounds, scents belonging to these objects; of boundless space in which they have their existence, and of an ever-rolling stream of time bringing change and incident. It tells of other life than mine busy about its own purposes.

As a scientist I have become mistrustful of this story. In many instances it has become clear that things are not what they seem to be. According to the story teller I have now in front of me a substantial desk; but I have learned from physics that the desk is not at all the continuous substance that it is supposed to be in the story. It is a host of tiny electric charges darting hither and thither with inconceivable velocity. Instead of being solid substance my desk is more like a swarm of gnats.

So I have come to realise that I must not put overmuch confidence in the story teller who lives in my mind. On the other hand, it would not do to ignore him altogether, since his story generally has some foundation of truth more especially in those anecdotes that concern me intimately. For I am given a part in the story, and if I do not take my cue with the other actors it is the worse for me. For example, there suddenly enters into the story a motor car coming rapidly towards the actor identified with myself. As a scientist I cavil at many of the particulars given by the story teller—the substantiality, the colour, the rapidly increasing size of the object approaching—but I accept his suggestion that it is wisest to jump out of the way.

There are ponderous treatises on my shelves which tell another story of the world around me. We call this the scientific story. One of our first tasks must be to try to understand the relation between the familiar story and the scientific story of what is happening around us.

At one time there was no very profound difference between the two versions. The scientist accepted the familiar story in its main outline; only he corrected a few facts here and there, and elaborated a few details. But latterly the familiar story and the scientific story have diverged more and more widely —until it has become hard to recognise that they have anything in common. Not content with upsetting fundamentally our ideas of material substance, physics has played strange pranks with our conceptions of space and time. Even causality has undergone transformation. Physical science now deliberately aims at presenting a new version of the story of our experience from the very beginning, rejecting the familiar story as too erratic a foundation.

But although we try to make a clean start, rejecting instinctive or traditional interpretations of experience and accepting only the kind of knowledge which can be inferred by strictly scientific methods, we cannot cut ourselves loose altogether from the familiar story teller. We lay down the principle that he is always to be mistrusted; but we cannot do without him in science. What I mean is this: we rig up some delicate physical experiment with galvanometers, micrometers, etc., specially designed to eliminate the fallibility of human perceptions; but in the end we must trust to our perceptions to tell us the result of the experiment. Even if the apparatus is self-recording we employ our senses to read the records. So, having set the experiment going, we turn to the familiar story teller and say "Now put *that* into your story". He has perhaps just been telling us that the moon is about the size of a dinner plate, or something equally crude and unscientific; but at our interruption he breaks off to inform us that there is a spot of light coinciding with division No. 53 on the scale of our galvanometer. And this time we believe him—more or less. At any rate we use this information as the basis of our scientific conclusions. If we are to begin actually at the beginning we must inquire why we trust the story teller's information about galvanometers in spite of his general untrustworthiness. For presumably his fertile invention is quite capable of "embroidering" even a galvanometer.

I do not want to spend time over points which no scientifically-minded person disputes; so I will assume that you agree that the only channel of communication between the story teller who lives in your mind and the external world which his story professes to describe is the nervous system in your body. In so far as your familiar conception or picture of what is going on around you is founded on your sense of sight, it depends on impulses transmitted along the optic nerves which connect the retina with the brain. Similarly for your other sense organs. You do not, of course, perceive the impulses themselves; the story teller has worked them up into a vivid story. The inside of your head must be rather like a newspaper office. It is connected with the outside world by nerves which play the part of telegraph wires. Messages from the outside world arrive in code along these wires; the whole substratum of fact is contained in these code messages. Within the office they are made up into a presentable story, partly by legitimate use of accumulated experience but also with an admixture of journalistic imagination; and it is this free translation of the original messages that our consciousness becomes aware of.

If we had a complete record of the impulses transmitted along the nerves we should have all the material which the story teller can have had as a foundation for his story—in so far as his story relates to the external world. And it is to this material that we must appeal if we wish to discover the truth behind the story. To appreciate the task of physical science let us then suppose that we are in possession of these data—the dots and dashes, or whatever the signals are, that arrive at the brain cells at the terminations of the nerves. All that physical science can assert about the external world must be inferable from these. If there is any part of our conception of the physical universe which cannot have come to us in the form of nerve signals we must cut it out. As in a beleaguered city there spread circumstantial rumours of happenings in the world outside which cannot have been received from without, so in our minds there arise all sorts of conceptions of entities and phenomena in the external world which cannot have been transmitted to us from outside. They do not conform to the type of message which the narrow threads of communication will bear. We are continually making the mistake of the man who, on receiving a telegram, thinks that the handwriting is that of the sender. The messages as we become aware of them in consciousness are dressed up with conceptions of colour, spatiousness, substance. This dress is no part of the message as it was handed in by the external universe. It is assumed after the message arrives; for the transmitting mechanism is by its very nature incapable of conveying such forms of conception.

This limitation of the transmitting mechanism is strikingly illustrated when we talk with a colour-blind person. We know from his amazing mistakes that there is a big difference between his perception of his surroundings and ours. But he is quite unable to convey to us how his perception differs. When he confuses red with green, does he see both colours as red or both as green or as some hue unknown to us? He has no means of telling us. The intrinsic nature of his perception is trapped in his mind. It cannot flow out along his nerves; nor could it travel up our nervous system if it reached it. Similarly the sensory qualities of colour, sound and scent cannot have been transmitted to us from the object in the external world to which we attribute the colour, sound and scent; for even if we suppose the object itself to be endowed with such qualities it would be as impotent as the colour-blind person to convey to us their character. The part played by the external object is to condition directly or indirectly the signals which pass along the nerves. The story which arises in our consciousness is a consequence of these signals, but it contains much that does not belong to the external message.

The inference of any kind of knowledge of the physical objects which lie at the far end of these lines of communication must evidently be very indirect. In this respect it differs from the knowledge constituted by the mind's immediate awareness of its own sensations, thoughts, emotions. I have elsewhere expressed this in the words: "Let us not forget that mind is the first and most direct thing in our experience; all else is remote inference".* That is a statement which, I believe, physicists accept almost as a truism, and philosophers generally condemn as a hoary fallacy. It is difficult to understand why there should be such a difference between us. I had thought that, like many other differences, it might arise because we do not talk the same language; but some recent writings seem to show that the cleavage may be deeper, and that there is a tendency in modern philosophy to adopt a view which is scientifically untenable.†

* *Science and the Unseen World*, p. 24. † See pp. 280–288.

Scientific thinkers generally agree that the channel of communication between the external world and man's consciousness is severely limited in this way; but, whilst giving intellectual assent, they do not always adjust their scientific outlook to correspond. They are strangely reluctant to doubt the assertions of the familiar story teller even when it is evident that he is talking through his hat. The feeling that many of the conclusions of relativity theory and quantum theory are contrary to common sense is largely due to this tenacity. We cling to certain features in the familiar picture of the external world, almost as though we were persuaded that some part of our percipient selves had been projected outside the body, and had entered into external things and become aware of their ultimate nature in the same direct way that the mind is aware of its thoughts and sensations. We uphold the familiar conceptions of space in the external world as assuredly as if the spirit of man could enter into space and feel what it is like to be large or small. But when an external object raps on the door at the extremity of a nerve, you cannot put your head outside to see what is rapping. You cannot know more of its nature than that it must be such as to account for the delivery of the raps in their sequence. A scientific theory which accounts for the raps is none the worse because it runs counter to the story teller's habitual but unwarranted picture of what lies beyond the ever-sealed door.

II

Broadly speaking the task of physical science is to infer knowledge of external objects from a set of signals passing along our nerves. But that rather underrates the difficulty of the problem. The material from which we have to make our inferences is not the signals themselves, but a fanciful story which has been in some way based on them. It is as though we were asked to decode a cipher message and were given, not the cipher itself, but a mistranslation of it made by a clumsy amateur.

It is true that the physiologist nowadays is able to tap the messages as they pass along a nerve. He can record the changes of electrical potential that occur when a nerve is stimulated, and the record shows a series of oscillations which are presumably the physical foundation for the perception that arises in the mind. But we cannot begin the study of the external world with these records. In order to utilise them a rather advanced scientific knowledge of the nature of the human body and the functions of the various nerves is presupposed. All that the physiologist has done is to tap the messages on the way to one brain and divert them into another brain—his own. That is not fundamentally different from the method of the physicist who intercepts the messages emanating from physical objects before they reach any nerve, and, for example, causes them to record themselves on a photographic plate. By one route or another the messages must ultimately be conducted to a seat of consciousness if they are to be translated into knowledge.

It is the inexorable law of our acquaintance with the external world that that which is presented for knowing becomes transformed in the process of knowing.

Thus in saying that the initial data of physics are nerve signals, we must not be confused by the fact that nerve signals are pictured by us as known processes in the external world. This identification of our initial data is not itself an initial datum; it is one of our indirect inferences. It all emphasises the difficulty of tracing our knowledge of the physical world to its beginning. We detect it stealing into our minds through our nerves; but our knowledge of the physical world had to be considerably advanced before we discovered that we possessed a nervous system.

More by the exigencies of its own development than by the considerations that we have been discussing, modern physics has been forced to recognise the gulf between the external world which appears in the familiar story of perception and the external world which presents its messages at the door of the mind. It is for this reason that the scientific story is no longer a tinkering of the familiar story but follows its own plan. I think the modern view can best be expressed by saying that we treat the familiar story as a cryptogram.

Our sensory experience forms a cryptogram, and the scientist is a Baconian enthusiast engaged in deciphering the cryptogram. The story teller in our consciousness relates a drama—let us say, the *Tragedy of Hamlet*. So far as the drama is concerned the scientist is a bored spectator; he knows the unreliability of these play-writers. Nevertheless he follows the play attentively, keenly alert for the scraps of cipher that it contains; for this cipher, if he can unravel it, will reveal a real historical truth. Perhaps the parallel is closer than I originally intended. Perhaps the *Tragedy of Hamlet* is not solely a device for concealing a cryptogram. I would admit —nay, rather I would insist—that consciousness with its strange imaginings has some business in hand beyond the comprehension of the cipher expert. In the truest sense the cipher is secondary to the play, not the play to the cipher. But it is not our business here to contemplate those attributes of the human spirit which transcend the material world. We are discussing the external world of physics whose influences only reach us by signals along the nerve fibres; and so we have to deal with the story after the manner of a cryptogram.

The solution of a cryptogram is found by studying the *recurrency* of the various signs and indications. I do not think we should ever have made progress with the problem of inference from our sensory experience, and theoretical physics would never have originated, if it were not that certain regularities and recurrencies are noticeable in sensory experience. We call these regularities of experience laws of Nature. When such a law has been established it becomes also a rule of inference, so that it helps us in further decipherment just as in solving an ordinary cryptogram.

I do not know how a logician would classify the process of solving a cryptogram. The decoded message is inferred from the cryptogram, but the method of inference can scarcely be described as logical deduction. In saying that the scientific description of the external world is inferred from our sensory experience, and that the entities of the physical world are *inferences*, I use the word inference in this broad sense.

Our task then is to discover a scheme revealed by the regularities and recurrencies in our sensory experience. Since these regularities occur in the sensory experience of all men the scheme is presented as an *external* world linking together the experiences of different individual consciousnesses. In thus defining the object of our search we determine to a certain extent the nature of that which we shall find. The universe of physics must by its very definition have the two characteristics of regularity (or partial regularity) and externality. We do not contest the right of anyone who is interested in other aspects of sense data, or of the consciousness in which they reside, to pursue his investigations in his own way; but so far as physical science is concerned we drop everything that is inessential to the elucidation of regularities and recurrencies.

I must also emphasise the significance of the term "external". The familiar world of my perception seems to be external; but, in the courts of science, what the familiar story teller says is not evidence. The world of my dreams also seems to be external, but it has no existence outside my mind. The argument that the world containing the entities of physics is external is quite independent. When I examine the content of my consciousness with a view to formulating the recurrencies of my sensory experience, there are two possible ways of treating the data—two ways in which I might attempt to solve the cryptogram. Among the data are certain auditory sensations "spoken words" and certain visual sensations "printed words" which admit of alternative treatment. I might study their recurrencies and regularities without discriminating them from other auditory and visual sensations. Then all the recurrencies are of data within my own consciousness and the study of them never takes me outside the region of my own mind; the solution of the cryptogram, if any, reached by this treatment will be an internal egocentric world. But such a treatment of the problem of experience is not often promulgated—if only because a lecturer cannot deny himself the hope that his "spoken words" will be treated by his audience as on a different epistemological footing from the beating of a tin can. Therefore in science and in most philosophies spoken and printed words are treated, not only as immediate sensory data of our own consciousness, but as communicating to us data existing in other consciousnesses.

Thus our first intimation of externality has no direct connection with physical science. It comes from the recognition that the problem of experience is concerned with data distributed among many different individual consciousnesses. The synthesis of experience then necessarily leads to the contemplation of a neutral domain not coextensive with any individual mind. Thus although we start from individual mental data, as soon as we commit ourselves to the recognition of other minds than our own, we are led to the conception of an external domain (physical space and time) to contain the inferential objects of our combined knowledge. Among these inferential objects are the nerve fibres and brain cells where (as the decipherment of the cryptogram progresses) the sources of communication between the objects of this external world and an individual consciousness are found to be located.

We asked why the story teller should be believed when he

talks about galvanometers, although he is untrustworthy when he talks of familiar objects. I think the answer is that the truth of the story is not the point in question; the physicist is concerned only with the scraps of cipher contained in it. The galvanometer is a device for leading the story into situations in which the underlying cipher becomes less baffling to interpret; it is not a bridle on the story teller's imagination.

III

A feature of progress in unravelling the cryptogram has been that much of our sense data proves to be redundant—redundant, that is to say, in the study of recurrencies. We can, for example, drop the sense of hearing, since it only indicates regularities which can alternatively be detected by our other senses. With the reduction of the number of types of sense data to a minimum there has been a parallel unification of the external world. One scheme of regularity suffices, instead of a distinct scheme for each of our senses, with perhaps additional schemes corresponding to electric and magnetic senses which we presumably might have possessed if Nature had so chosen. This dropping of a variety of types of sense data is responsible for some of the most striking differences between the familiar and the scientific conception of the external world.

Writing this chapter on an autumn day, I feel myself in a familiar world whose most prominent characteristic is colour. There is no colour in the physical world. I think that that is the right way to put it. It is true that each colour is represented in the physical world by a number supposed to indicate the length of a wave of some kind. Similarly I am represented at the telephone exchange by a number indicating a hole in a switch-board; but it would not be correct to say that I inhabit the telephone exchange. To put it another way, there is nothing in the accepted description of the physical world which owes its acceptance to the fact that we have a sense of colour. Everything that we assert can be verified by a colour-blind person; and indeed most of our accurate knowledge has been ascertained through the medium of a colour-blind photographic plate.

When we have eliminated all superfluous senses, what have we left? We can do without taste, smell, hearing, and even touch. We must keep our eyes—or rather one eye, for there is no need to use our faculty of stereoscopic vision. The eye need not have the power of measuring or graduating light and shade; I think it is sufficient if it can just discriminate two shades so as to detect whether an opaque object is in a certain position or not.

With this reduced equipment we can still recognise geometrical form and size. We can recognise that one object appears round and another square, or that one is apparently larger than another. Some years ago the position had been reached that spatial form and magnitude were the only features in our familiar picture which existed also in the external world of science. This led to a geometrisation of physics. You can see why at this stage physics became so largely geometrical in its methods and vocabulary. The preserved data which contained the recurrencies, and therefore the key to the cryptogram, were wholly geometrical; all other data had been dropped as redundant when it was found that they revealed only the same recurrencies as the geometrical data.

By limiting the sensory equipment of our observers we do a great deal to stop their quarrelling. For example, by removing their ears we put an end to the disputes of the musical critics. I do not say that they are disputing about nothing; but, whatever it is, it is not relevant to the scheme of regularities of which the physicist is in search.

But it was found that the observers were still quarrelling

SCIENCE AND EXPERIENCE

even when they had only form and size to quarrel over.*
So in 1915 Einstein made another raid on their sensory
equipment. He removed all the retina of the eye except one
small patch. The observer could no longer recognise form
or extension in the external world, but he could tell whether
two things were in apparent coincidence or not.

If you read about Einstein's theory of relativity you will
find many references to a peculiar person called "the observer"—the man who has a habit of falling down lifts, or
getting transported by aeroplanes travelling at 161,000 miles
a second. Now you have a picture of him. He has one eye
(his only sense organ) which is colour-blind. He can distinguish only two shades of light and darkness so that the
world to him is like a picture in black and white. The sensitive
part of his retina is so limited that he can see in only one direction
at a time. We allow him any number of assistants equipped
like himself so that they can keep watch on the different
parts of an experiment and pool their knowledge afterwards.
Since we have so mutilated him he cannot make the experiments himself. We perform the experiments, and let him
keep watch. The point is that all our knowledge of the
external world as it is conceived to-day in physics can be
demonstrated to him. If we cannot convince *him* we have
no right to assert it.

I will not stop to justify in detail this drastic method of
inculcating respect for truth. I will only point out that it is
not too intrinsically absurd, because we have left the observer
power to recognise that a pointer coincides with a graduation
on a scale. Practically every physical measurement which
has any pretension to accuracy resolves itself into a pointer-
reading of this kind. Instead of relying on our sense of
warmth we read the graduation of a thermometer; instead
of using our inner feeling of duration we read the dial of a
clock. Thus the observer will generally have no difficulty

* See, for example, *The Nature of the Physical World*, pp. 12-16.

in deciding questions of exactitude. His mutilation will make it rather difficult for him to keep general track of what the experiment is concerned with; but by the aid of the army of assistants that we have allowed him, he will be able to maintain a sufficient watch on all parts of the apparatus.

I will give one example to show how in scientific practice pointer-readings are substituted for diverse sensory data. Our ideal observer is supposed to have no sense of the graduation of light and shade; therefore when he looks up at the night sky all the starry points will look to him alike in brilliance. Will not this rather disqualify him as an astronomer? Not at all. For let us consider how in practice a professional astronomer recognises the differences of brightness of the stars. It happens that this is the work with which my own observatory is now chiefly occupied. We follow a method (used also in a few other places) which for the brighter stars is found to give by far the most accurate results. The light of the star is concentrated by a telescope so as to enter a photo-electric cell. But first, how do we know that we have got hold of the right star—that we can recognise again the star which we have been measuring? Stars are commonly recognised by the patterns that they form with other stars—crosses, triangles, W's, etc.; but it will be remembered that Einstein has cut down the field of vision of our ideal observer so that he cannot see these patterns. No matter. The observer at Cambridge would in any case be unable to see the patterns, because the telescope is so constructed that the observations are made in a closed room without a glimpse of the sky; and when the photo-electric apparatus is mounted, the observer cannot see through the telescope more than one star at a time—just as though Einstein had really operated on his retina. The star is set for and identified by reading two graduated circles attached to the telescope. Thus, even in the identification of the star, pointer-readings are substituted for other sensory data.

SCIENCE AND EXPERIENCE

The light on reaching the photo-electric cell liberates electrons from a film of potassium, and these are driven by a constant electromotive force (which incidentally is measured by another pointer-reading, viz. that of a voltmeter) on to the needle of an electrometer. Omitting technical details the task of the observer is to watch the pointer-needle of the electrometer travel from coincidence with one graduation of a scale to coincidence with another graduation, timing it with a stop-watch. The stronger the light of the star, the faster the passage. So that finally the determination of the brightness of the star resolves itself into yet another pointer-reading, namely that of the hand of a stop-watch on its graduated dial.*

"One star differeth from another star in glory" wrote the apostle. The Nautical Almanac is more precise: 2 Ceti, $4^{m\cdot}62$; α Andromedae, $2^{m\cdot}15$; β Cassiopeiae, $2^{m\cdot}42$; and so on. Even the glory of the sun has been systematised in the same way as $-26^{m\cdot}7$ on the scale of magnitudes. "How art thou fallen from heaven, O Lucifer, son of the morning!" All thy glory has been turned into the pointer-reading of a terrestrial stop-watch.

IV

If a catalogue of pointer-readings were the ultimate end, we might well question whether physical truth were worth the seeking. But the pointer-readings are rather the beginning, replacing the story teller's romances which from our point of view must be looked on as a false start. They constitute the material which contains all the recurrencies whereby the cryptogram is deciphered, since we find by experience that the use of a wider variety of sensory data only leads to

* The pressing of the stop-watch at the right moment involves a sense of touch, so that in this respect the Cambridge observers fall short of our theoretical ideal. But the principle would not be affected if an automatic method of timing the motion of the needle were substituted.

redundancies. In later chapters we shall learn some of the results of the deciphering; and perhaps you will be persuaded that the reconstructed story of the stars is a not inadequate compensation for setting aside the familiar story teller's romantic imaginations.

But, it may be suggested, if all observation is reduced to coincidences and pointer-readings, can we ever infer from it anything but a system of relationship of coincidences and pointer-readings? In one sense the answer is No. But if the question is put in the form "Can we by manipulating pointer-readings ever arrive at a knowledge which does not smell of pointer-readings?" I suggest that it might equally well be asked "Can an artist by manipulating paint ever achieve a creation which does not smell of paint?" But I do not wish to set this question aside lightly, for it goes to the very heart of the difference between the new and the old scientific outlook. We shall see later that a scheme of relationship, or a *structure*, has a significance which can be abstracted from the intrinsic nature of that which is the subject of the relationship. The structure is the object of our search, and when we have reached knowledge of the structure we can disregard the scaffolding by which we reached it. It does not lessen the dignity of the structure that its elements are pointer-readings—which after all is only the story teller's name for them.

If none of the images which constitute our sensory perception are applicable to the physical world, in what form can our knowledge of the physical world be expressed? We have deciphered our cryptogram, but the result is a message couched in unknown language which we have no hope of translating into the language of the story teller. It does not, however, follow that it is unintelligible to the mind. Perception is only part of our mental outfit, and the language of perceiving is only part of the language of knowing. Our reading of the cipher of experience leads to an understanding of our environment, highly abstract indeed and only to be apprehended by the intellect through symbolic expression, but nevertheless satisfying to the urge of the human spirit in its quest for knowledge. In Chapter XII I will try to explain in some detail how a genuine knowledge of the external world can be expressed and apprehended without referring to perceptual images. Here I will content myself with one illustration.

The sentence which I am now writing can exist in a number of forms. It may be a series of sounds perceived by a listener. It may be printed in a book. It may be recorded by a gramophone, and exist as a trace on a disc. It was originally a mental composition unuttered and unwritten. There is something common to all these forms; and that common element, if we can abstract it, constitutes the sentence.* There may well be forms of existence of a sentence which are unimaginable to us to-day, just as a hundred years ago it could scarcely have been imagined that a sentence could exist as a gramophone record. The various forms are described in terms of familiar images—sounds, discs, black and white shapes—but the sentence itself is detached from all familiar images. (I would again remind you that I am referring to the exact words, not to the meaning.) That does not render our knowledge of the sentence unsatisfying or incomplete. In telling a child of Nelson's famous message to the Fleet, it is not necessary to prefix a discourse on the methods of naval signalling. And if we could foresee that a hundred years hence a certain sentence would pass from one individual to another, that would be precise and intelligible knowledge of the future, notwithstanding that the transmission might be by methods as yet unimagined by us and therefore unspecified in terms of familiar images.

The sentence which constitutes the solution of an ordinary

* I am not referring to the meaning, which might be conveyed by a different sentence or in a different language.

cryptogram is not associated with any one form of existence. Likewise the external world of physics which is the solution of the cryptogram of our sensory experience is not associated with any one form of existence. This means that when we consider experience as a whole, in passing from the mental experience to the phenomena of the physical world we do not encounter any discontinuity in the form of existence, unless we deliberately create a discontinuity.* There is a difference, of course—for the object of our analysis is to differentiate—but not a dualism. The older philosophic dualism of mind and matter seems to have been that of the man who has received one part of his instructions verbally and the other part in written form and feels unable to combine them because of the incompatible nature of sound waves and ink.

By the dropping of redundant sense data we have reduced our observational material to pointer-readings, or more generally to coincidences. Einstein's general theory of relativity (1915) was based on the principle that observable data are always describable by coincidences, or, as the favourite expression was, "intersections of world-lines". Clearly any inference we draw, any structure which we ascribe to the external world, must be of such a character that it is invariant for any changes which we may make in our picture which do not alter these intersections of world-lines, i.e. turn intersections into non-intersections. Our inference has to have a fluid form. If we conceive a framework of lines whose intersections correspond to the observed coincidences, then however the framework is distorted and twisted it will still represent all that we can really know, provided that the joints are not tampered with. I suppose that a musician who listens to a broadcast performance can see in his mind the movements of the fingers and even the

---

* Just as we may create a discontinuity of form between a cryptogram and its solution by giving one in written form and the other orally.

swaying of the head and body of the pianist; but setting aside his preconceptions all that he can really infer from the sound is that certain keys have been struck with greater or less force for longer or shorter times, and any scheme of movement leading to this result is an equally admissible inference. In the same way, setting aside preconceptions, we cannot discriminate between the various possible systems of structure of the external world which would lead to the same sequence of impulses on the extremities of human nerves; or since the structure of the nervous system is itself a matter of inference, we can transform the whole structure of the physical world (nerves included) in any way which does not alter the sequence of impulses reaching those points in the structure identified as doors of communication with consciousness. The solution of the cryptogram has (like the sentence) many forms of existence, and also (unlike the sentence) many equivalent and equally admissible representations within the same form of existence.

It is this fluidity of representation—so different from the representation of our environment in the story teller's version—which first found its way into physics in Einstein's theory of relativity. That is why the theory of relativity is such an epoch-making breach with tradition. It is interesting to notice that this revolution of thought had birth within physics itself. I have been arguing that from the very nature of our acquaintance with the physical world there must necessarily be a fluidity of representation of that which we discover about it—that many apparently different representations of the world-structure are equivalent in all that concerns observation. But it was not by this kind of reasoning that the question first arose in physics. The physicist in the ordinary course of his work had stumbled upon a multiplicity of representation. He was very much bothered by it. He thought it was his duty to decide which representation was the "right" one. There were things in Nature which he had never doubted were quite definite; the story teller said so, and that was good enough for a man who dealt with hard facts. Yet Nature by the most artful devices persistently refused to disclose anything definite about them. I do not mean that Nature is characteristically indefinite and slovenly; but she is definite in her own way, not in the story teller's way. At last the physicist was forced by his own discoveries to consider more philosophically the principles of knowledge and the kind of truth that his methods were adapted to ascertain.

V

At present theoretical physics is sharply divided into macroscopic theory and microscopic theory, the former dealing with systems on a scale perceptible to our gross senses and the latter with the minute atomic substructure underlying the gross phenomena. Broadly speaking, macroscopic systems are covered by relativity theory and microscopic systems by quantum theory. The two theories must ultimately be amalgamated; the amalgamation is in fact now in progress. But for the purposes of a general survey it is easier to think of them as distinct.

Microscopic physics introduces entities—molecules, atoms, electrons, protons, photons, etc.—which do not appear at all in the familiar story. There has been a tendency among scientific philosophers to regard these as having a more hypothetical status than the objects studied in macroscopic physics. Prof. H. Dingle's *Science and Human Experience* is a typical example of this attitude. According to him atoms and electrons are unverifiable hypotheses, "existences whose unobservability is part of their essential nature" (p. 47). He is contrasting them with ordinary "observable" objects, and he intends to convey that they have not the same kind of connection with human experience as the more ancient denizens of the physical world such as sticks and stones and stars. This distinction appears to me quite unwarranted.

An electron is no more (and no less) hypothetical than a star. Nowadays we count electrons one by one in a Geiger counter, as we count the stars one by one on a photographic plate. In what sense can an electron be called more unobservable than a star? I am not sure whether I ought to say that I have seen an electron; but I have just the same doubt whether I have seen a star. If I have seen one, I have seen the other. I have seen a small disc of light surrounded by diffraction rings which has not the least resemblance to what a star is supposed to be; but the name "star" is given to the object in the physical world which some hundreds of years ago started a chain of causation which has resulted in this particular light-pattern. Similarly in a Wilson expansion chamber I have seen a trail not in the least resembling what an electron is supposed to be; but the name "electron" is given to the object in the physical world which has caused this trail to appear.* How can it possibly be maintained that a hypothesis is introduced in one case and not in the other?

Thus when we discuss the reality of the physical world and the entities which constitute it, we have no reason to discriminate between the macroscopic and the microscopic entities. It is to be treated as a whole. If the physical world is a hypothesis, stars and electrons are hypothetical; if the physical world is an inference, stars and electrons are inferential; if the physical world exists, stars and electrons are real. Of course we must not forget that science is progressing, and that the various entities now regarded as composing the physical world are, as it were, on probation. But this domestic uncertainty within the scientific scheme is not here a point at issue. It is the principles of physical science rather than the interim results which we are examining critically in this chapter. Perhaps we may usefully borrow a phrase

* See Plate 1.

from commerce and finance. The letters, E. & O.E., in a document stand, I am told, for "errors and omissions excepted". My contention is that atoms, electrons, and other entities of microscopic physics (E. & O.E.) are hypotheses, inferences or realities according as chairs and tables and other commonplace objects of the physical world are hypotheses, inferences or realities.

When we stripped our ideal observer of most of his sense organs we left him part of an eye in order that he might observe coincidences. Was not this a rather arbitrary selection from among his diverse sense organs? Perhaps it was. It was enforced entirely by practical considerations; I will not defend it on philosophic grounds. I will not enter into an argument with my dog as to whether the eye or the nose is ideally the more trustworthy organ for exploring the external world. All I assert is that in a competition between various observers, each allowed only one kind of sensory impression, the Einstein observer has up to the present gone furthest in discovering the scheme of regularity underlying all sensory impressions. The technique of the practical physicist has come more and more to depend on observing coincidences (pointer-readings and similar measurements). Inferences from our other perceptions partially reveal the same scheme of regularity in Nature, but they do not go so far in unravelling it. (The unity of the scheme underlying all our diverse perceptions is not an a priori judgment; it is a conclusion, possibly mistaken, which we have drawn from such fragments of the scheme as have already been discovered.) As the advantage appears to be a purely practical one, I do not think we should be justified in attributing a special philosophic importance to the perception of coincidences. In particular we ought not to displace our mental image of coincidence into the external world.

In the earlier scientific outlook we used to suppose that shape and size existed in the external world precisely as they

SCIENCE AND EXPERIENCE 23

appear in our perceptions—not like colour which had to be represented by a wave-length. Perhaps most physicists would now transfer coincidences in the same way, and suppose that the coincidences and intersections referred to in the scientific description are just like the coincidences and intersections in our mental picture. I do not think that this naïve displacement of essentially mental forms of relation is permissible; and it is interesting to notice that the quantum theory gives a distinct warning against it.

The observed coincidences of gross matter are, of course, only approximate contacts; but as we deal with smaller and smaller particles the conception of coincidence can be refined to higher and higher exactitude. If coincidence were the key-stone of world structure we should expect to find the greatest refinement of it in the theory of atoms and electrons. But on the contrary modern physics represents atoms and electrons in a scheme which *forbids* coincidence. There is a fundamental law called Pauli's Exclusion Principle which asserts that two electrons can never be in the same cell of the phase-space in which we represent them.

In the quantum theory we abandon the last vestige of any displacement of the elements of the familiar world into the physical world. The connection is not displacement but inference. The inferences do not resemble the sense data any more than criminals resemble clues. I have hesitated whether I ought not to make one reservation. We displace integers freely from the familiar world to the physical world. An apple in the familiar story has a counterpart in the external world; none of our familiar conceptions are appropriate to describe the nature of this counterpart, and we can only indicate it by a symbol such as X. But at any rate we can then say that the counterpart of two apples in the familiar story is two X's. I grant that; but I would not like to commit myself to the opinion that the *twoness* of two X's is just like the twoness of two apples. In the case of electrons I would go further; I do not believe that the twoness of two electrons is a bit like the twoness of the two apples in the familiar story. In fact multiplicity in the external world should be regarded as a property (indescribable in familiar terms) which, being by its nature discontinuous, has been correlated to the series of arithmetical integers, just as continuous properties are correlated to continuous measure numbers.

VI

In view of the closer contact which now exists between science and philosophy, I would like to raise one question which affects our cooperation. A feature of science is its *progressive* approach to truth. Is there anything corresponding to this in philosophy? Does philosophy recognise and give appropriate status to that which is not pure truth but is on the way to truth? Let me here warn the reader that, whilst in this opening chapter I set before him the ideal after which we strive in sifting the truth about the external world from the imaginations of the familiar story teller, I shall not keep to the same austere outlook in subsequent chapters. Indeed it would lack a sense of proportion to use the steam-hammer of critical philosophy to crack every nut on the tree of science. It is essential that philosophers should recognise that in dealing with the scientific conception of the universe they are dealing with a slowly evolving scheme. I do not mean simply that they should use it with caution because of its lack of finality; my point is that a vehicle of progress is not furnished on the same lines as a mansion of residence. The scientific aim is necessarily somewhat different from the philosophic aim, and I am not willing to concede that it is a less worthy aim.

It would be no aid to science if philosophers enforced on us their glimpses of pure truth centuries before our scheme was ripe to receive it. Perhaps if it had been guided by philosophy, physics would have been relativistic long before Einstein; but I feel sure that physics would not now have been in so advanced a state if it had never passed through the non-relativistic phase of the nineteenth century. So when after laborious research physics arrives at "revolutionary conclusions" which philosophy claims to have known from its cradle, there are two versions to the story. According to one the physicist is a workman of pig-headed disposition who would have got on much faster if he had listened to the advice of philosophers. The other is that the philosopher is an officious spectator who bothers the workman by handing him tools before he is ready to use them. I daresay that, as is usual in such cases, the truth lies somewhere between the two versions.

I suppose that before concluding I must encounter the plain question, Does the external world described in physics (E. & O.E.) really exist? But I do not consider it to be a "plain question". The difficulty is that the words *existence* and *reality* require definition like any other terms that we employ, and I do not know where to turn for a recognised definition. There is no reason to think that a physicist, a mathematician and a philosopher attach the same meaning to the word "existence". Descartes seems to have believed that he existed because he thought. Dr Johnson seems to have believed that the stone existed because it was kickable. Others have regarded their own existence as a debatable question.

For my part, any notion that I have of existing is derived from my own existence; so that my own existence is a tautological consequence of any definition that I should be willing to adopt. Other conscious beings also exist, for I am convinced that I must not deny to them the attributes I recognise in myself. I thus lay down the rudiments of a "web of existence" to which all that enters into knowledge is related in various ways. I have tried to show the particular way in which the world of physics is related. I expect that most people would regard a world related in this way as thereby qualified to be considered part of the same web of existence; but I cannot feel any great interest in this desire to employ a vague instead of an exact description of the relation. However, so far as I can judge the meaning of the question, the answer appears to be in the affirmative—the external world described in physics (E. & O.E.) really exists.

One thing can perhaps usefully be added. I do not think that with any legitimate usage of the word it can be said that the external world of physics is the *only* world that really exists.

CHAPTER II

DRAMATIS PERSONAE

> These our actors,
> As I foretold you, were all spirits, and
> Are melted into air, into thin air.
> SHAKESPEARE, *The Tempest*.

I

IT is frequently necessary in the following chapters to refer to the chief results of atomic physics and to our general knowledge of atoms, radiation and aether. Many excellent nontechnical books on the subject are available, and I do not wish to linger over a fascinating but oft-told story. It has, however, seemed desirable to include here a brief review of our knowledge.

We have seen that the ultimate scientific description of the physical universe must be divorced from all familiar images; but here we shall follow the working conceptions of the experimental physicist rather than those of the extreme theorist. Scientific conceptions relate to a number of different levels, and we do not need to call up the ideas of the profoundest level for every minor occasion. It would be inappropriate to think in terms of atomic theory in the act of stepping off a bus; and similarly the physicist who splits atoms may, in a practical sense, quite well understand what he is doing without invoking the more recondite conceptions of wave mechanics or of the theory of groups. So in this chapter I do not at first descend to the foundations, but halt at a level which is important because it has supplied a great deal of the current vocabulary of physics. My description cannot attempt greater accuracy or profundity than that of the level to which it belongs.

It appears that all matter is constructed from two kinds of elementary particles called protons and electrons. The proton carries a certain definite charge of positive electricity and the electron an equal charge of negative electricity. But these two kinds of particle are not in all respects the exact opposite of one another; for the proton is very much heavier than the electron, its mass being about 1847 times as great.

The true opposite of the electron was discovered about two years ago; it is called the positive electron or positron. It is a particle of just the same mass as an electron but with a positive instead of a negative unit charge. Apparently, however, positrons have only a momentary existence. They are created during certain kinds of intense discharge of energy—when cosmic rays fall on matter, or when an atomic nucleus is bombarded by fast-moving particles. But after travelling a short distance they vanish, having encountered ordinary (negative) electrons with the result that mutual cancellation takes place. Presumably the proton also has its opposite, a negative proton or negatron, but this has not yet been discovered.

Plate 1 shows the tracks of electrons and positrons, rendered visible by Prof. C. T. R. Wilson's method which causes small drops of water to condense along the tracks. The photograph, due to Blackett and Occhialini, shows a shower of these particles produced by a single cosmic ray falling on copper. A magnetic field was so placed that the electron tracks curve to the left and the positron tracks to the right. Most of the particles were going too fast to be much deflected, and therefore cannot easily be discriminated; but one positron is obvious, and another with smaller curvature is fairly evident.

Both protons and electrons must be pictured as exceedingly small, very much smaller than an atom. Formerly an electron was supposed to have a radius of $2 \cdot 10^{-13}$ cm. and the proton was supposed to be a great deal smaller; but now we regard

PLATE I

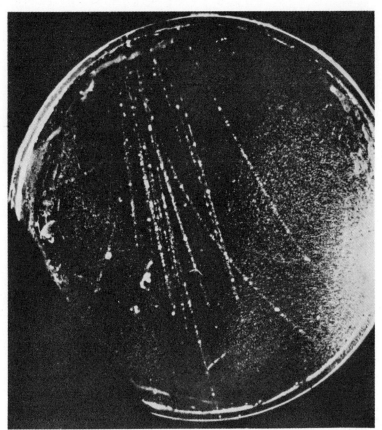

Blackett and Occhialini

ELECTRONS AND POSITRONS

The tracks pass downwards through a magnetic field which deflects electrons to the left, positrons to the right. One positron track with pronounced curvature to the right is easily distinguished. Two electrons are seen on the left of the photograph.

them both as mathematical points. This is not so much a correction of the original estimates of size as a recognition that the ordinary notions of space break down in the branch of physics which deals with these particles. It is found to be inappropriate to attribute extension to an electron, though we have, as it were, to make it up to the electron in other ways.

The first step in the construction of matter out of protons and electrons is the building of an atom. It is clear that any permanent structure must consist of an equal number of protons and electrons. For if there were an excess of protons there would be a net positive charge; and this would attract any negatively charged electrons in the neighbourhood and draw them into the structure until the excess had been neutralised. Although the protons and electrons in an atom are equal in number, there is great asymmetry in their arrangement. All the protons and about half the electrons are welded into a structure about 10^{-12} cm. in radius called the nucleus; the rest of the electrons, called satellite electrons, travel round the nucleus in relatively distant orbits, so that the whole atom extends to a radius of about 10^{-8} cm. The proportion is nearly the same as that of the sun and its planetary system; the sun, corresponding to the atomic nucleus, has a radius of 430,000 miles, whilst the limits of the solar system defined by the orbit of Pluto extend to 3,600,000,000 miles. We may thus picture an atom as a miniature solar system.

We can specify the different kinds of atoms by giving (1) the number of protons in the nucleus, and (2) the number of satellite electrons or, what comes to the same thing, the excess of the number of protons over the number of electrons in the nucleus. The first number gives approximately the mass of the atom (taking the mass of a proton as 1), for the electrons are so light that their masses scarcely count. The second number is called the atomic number; it gives the net (positive) charge of the nucleus. The chemical name of an atom is decided by the atomic number alone. For example, an atom with net nuclear charge 17, so that there are 17 satellite electrons, is called chlorine. But there are two common kinds of chlorine, one with 35 protons and 18 electrons in the nucleus and therefore of atomic weight 35, and the other with 37 protons and 20 electrons in the nucleus and therefore of atomic weight 37. We speak of these as two "isotopes" of chlorine. There are not many phenomena in which it makes much difference whether a 35-mass or a 37-mass chlorine atom is involved. In diffusion experiments the former should behave rather more nimbly; but in general the differences are so slight that chemists continually worked with chlorine for 150 years without discovering that it was a mixture of two kinds of atoms.

The atomic numbers of the elements range from 1 for hydrogen up to 92 for uranium. Elements have been discovered occupying all but two of these numbers. Many of them have, like chlorine, two or more isotopes, so that the total number of different kinds of stable atom now known exceeds 240. In addition there are many short-lived radioactive atoms. It is not certain that 92 is the upper limit for an atomic number; in fact, if we admit jerry-built atoms which collapse after a few minutes, element No. 93 has recently been created artificially by Fermi.

Element No. 1 requires special reference. Its simplest form is the ordinary hydrogen atom which consists of a proton and a satellite electron. It thus differs from all other elements in having an elementary particle instead of a complex structure for its nucleus. This distinction is so important that it is sometimes advantageous to regard matter as being of two main varieties, namely hydrogen and not-hydrogen (pp. 147, 167).

Recently an isotope called "heavy hydrogen" has been discovered; this has a nucleus consisting of 2 protons and

1 electron, so that it is of atomic weight 2; the net nuclear charge is 1, and there is just one satellite electron as in ordinary hydrogen. By the usual rule the two isotopes would not be entitled to separate chemical names; but the circumstances are rather exceptional, and heavy hydrogen has been named deuterium (or by some writers, diplogen) and given a chemical symbol D. Its nucleus, when it occurs without the satellite electron, is called a deuton (or diplon). The difference between hydrogen and deuterium (of respective weights 1 and 2) is not such a trivial matter as the difference between most isotopes; and deuterium and its compounds have appreciably different properties from hydrogen and its compounds. Naturally the compound D_2O, or heavy water, has received special attention; it is 11 per cent. heavier than ordinary water (H_2O). A still heavier hydrogen of atomic weight 3 has also been discovered; its nuclei (tritons?) consist of 3 protons and 2 electrons.

Another recent discovery is the neutron. This appears to be a nucleus consisting of 1 proton and 1 electron, so that the net charge is zero and there are no satellite electrons. It is thus an element of atomic number 0. We might describe it as an isotope of *nothing*. From another point of view the neutron is a kind of collapsed hydrogen atom; both consist of a proton and an electron, the difference being that in the neutron they are held close together by nuclear binding and in the hydrogen atom more distantly by satellite binding. One of the questions we ask ourselves is whether hydrogen atoms ever spontaneously collapse into neutrons.*

Another very familiar particle is the α particle.† It is the nucleus of a helium atom (atomic number 2) consisting of

---

* According to some experimenters the mass of a neutron is rather greater than that of a hydrogen atom. If so, it contains more energy, so that its formation involves an absorption of energy and is not of the nature of a spontaneous collapse.

† A β particle is merely a fast-moving electron.

4 protons and 2 electrons. This appears to be a particularly stable combination, and it was formerly thought that within the more complex nuclei a large proportion of the protons and electrons are grouped as α particles. But later investigations of the structure of nuclei are adverse to this hypothesis. It now appears that each electron is bound to a proton so as to form a neutron; thus the nucleus can be treated as an assemblage of neutrons and protons.

The view is now often advocated that the neutron is a simple elementary particle, and that the proton is a complex body composed of a neutron and a positron. I do not think that this can be accepted as fundamentally true. Doubtless there are phenomena for which it is convenient to transpose the equation, neutron=proton+electron, into proton=neutron−electron, or, since "minus an electron" is equivalent to a positron, into proton=neutron+positron; but I do not think the suggestion can be allowed any deeper significance.

II

Both the system of satellite electrons and the nucleus itself can be modified or broken by sufficiently energetic disturbances from outside. It does not require much energy to detach the outermost of the satellite electrons; so that in the laboratory, and much more frequently in the stars, we may find atoms without their full quota of satellite electrons and therefore having a net positive charge. These incomplete atoms are called ions. Ionisation does not involve any permanent change in the atom—any "transmutation of the elements"—for as soon as the disturbed conditions subside the nuclear charge, which has been left intact, will attract to itself the number of satellite electrons needed to balance it.

The energy needed to bring about an alteration in the nucleus is of a much higher order. But the physicist has at his disposal a number of fast missiles—electrons, protons, neutrons, deutons, tritons, α particles—either projected naturally by the discharges of radio-active elements or speeded up artificially by applying a large electromotive force. Using a sort of machine-gun fire of these missiles, the experimenter is able now and then to hit a nucleus with sufficient energy for the missile to penetrate and change the constitution of the nucleus either by adhesion or disruption. The atom is then transmuted into a different element.

It often happens that the element first created by such a bombardment is unstable; so that after a certain short average life a rearrangement of the internal structure takes place and a particle of some kind is shot out. Thus the original transmutation is followed by a second spontaneous transmutation. The ordinary radio-active atoms, uranium, radium, thorium, actinium, are likewise unstable atoms, only they are comparatively long-lived. Their well-known spontaneous transmutations, which are generally accompanied by the discharge of α particles or electrons, are the aftermath of an evolution of complex unstable nuclei, which presumably occurred in the highly disturbed conditions in the interior of the sun some thousands of millions of years ago before the earth became separated from the solar mass.

The mass of a nucleus is not precisely equal to the sum of the masses of the protons and electrons composing it; it is always a little less. This mass-defect is of great importance because it indicates the energy of formation of the nucleus. Protons and electrons naturally tend to drop into a configuration of smallest possible energy; and their tendency to form nuclei evidently implies that by so packing themselves their total energy is less than when they are apart from one another. Thus in the formation of a nucleus energy is set free. Actually it is radiated away as high-frequency radiation or carried off as kinetic energy by high-speed particles discharged during the steps of the formation. It is well known that energy and mass are two aspects of the same entity, and when the energy departs the corresponding amount of mass also departs. Thus the mass-defect records how much energy has left the system.

III

The chemical, optical and magnetic properties of an atom are almost wholly conditioned by the structure of its satellite electron system. According to the level of ideas that we are now following these electrons describe fixed orbits about the nucleus. But it is necessary to insist more strongly than usual that what I am putting before you is a *model*—the Bohr model atom—because later I shall take you to a profounder level of representation in which the electron instead of being confined to a particular locality is distributed in a sort of probability haze all over the atom; and it requires a close study of the mathematical equations to see that the two kinds of representation have anything in common. It is doubtless disconcerting to read in one chapter that an electron is confined to its groove and cannot pass to another groove without a discontinuous jump, and in another chapter that an electron in an atom cannot be located anywhere in particular; but I suppose that we were once disconcerted to find the world in two hemispheres on p. 1 of an atlas and in Mercator's projection on p. 2.

In Bohr's model there are a limited number of orbits available for the electrons. These orbits are laid down by a peculiar system of laws given in quantum theory. It is as though the field surrounding a nucleus were traversed by a number of paths, and electrons roaming in the field were instructed to keep to the paths. The orbits are classified in groups. Starting from the nucleus there are 2 small circular orbits forming group K; then come 8 larger orbits (6 circular and 2 elliptic) forming group L; then 18 still larger orbits forming group M; and so on. Ideally the series of orbits continues up to the limit set by the size of the universe; but in practice the territory governed by an atomic nucleus is limited by the claims of adjacent nuclei. The larger groups of orbits are divided into subgroups corresponding to their eccentricities, some being circular and others more or less strongly elliptical.

It is a law that no two electrons may occupy the same orbit (Pauli's Exclusion Principle). When the atom is in a normal state of quiescence its satellite electrons take up the arrangement of minimum energy, which means that in general they fill the orbits which are closest to the attracting nucleus. But, bearing in mind that the electrons repel one another, the problem of finding the arrangement of minimum energy is not altogether simple; and when the number of satellite electrons is large, it often pays to fill the more eccentric orbits of a higher group rather than the circular orbits of a lower group. By studying these arrangements it has been found possible to explain in detail both the regularity and the apparent irregularities in the sequence of chemical properties shown in the periodic table of the elements.

Crudely expressed, the fundamental law of chemistry is that a satellite electron likes to belong to a complete group or subgroup; it hates to be the odd man out. Helium with 2 satellite electrons can just complete group K; neon with 10 satellite electrons can just complete groups K and L. Argon (18) completes groups K and L and has 8 electrons in group M; although this does not complete the group, it completes the most symmetrical subgroup of group M. These atoms are so self-satisfied that they form "inert gases" and refuse to enter into combination with other atoms. Adding one electron to each of these, we have lithium (3), sodium (11) and potassium (19); they accordingly have one electron over which must start a new group or subgroup. This unhappy electron is called the *valency electron*, and it is responsible for the chemical activity and alkaline nature of lithium, sodium and potassium. Taking similarly a step backwards, chlorine (17) has 7 electrons in its *M* group; these are as restive and dissatisfied as a party of 7 bridge-players. It would be an admirable arrangement for both sides if chlorine could borrow sodium's lonely electron to complete its group. The arrangement can be made, and the two atoms combine to form a molecule of common salt (NaCl).

Besides matter, which we dissect into protons and electrons, the other chief performer in the drama of physics is radiation. Radiation is the general name given to electromagnetic waves, or waves in the aether which is the continuous background between the protons and electrons. These waves may be of any length (from crest to crest) or equivalently of any frequency or pitch. One particular octave can stimulate our optic nerves, and within this range of frequency the electromagnetic waves constitute light. Other ranges of frequency have other characteristic manifestations. Arranged in order of diminishing wave-length and increasing frequency, the waves are classified roughly as Hertzian or broadcasting waves, infra-red or heat rays, light, ultra-violet or photographic rays, X rays, γ rays. If the primary cosmic rays are electromagnetic waves they are of still higher frequency than the γ rays, but it now seems more probable that they are high-speed particles.

We shall now consider briefly how atoms and radiation interact with one another. When there is energy straying round in the form of radiation, or when the atoms are jostling one another with energy derived from their high temperature, the satellite electrons will not necessarily occupy the orbits which correspond to minimum energy. The atom is then said to be "excited". But the atom cannot take up just any quantity of energy; the amount has to be that which will lift an electron from one orbit to another vacant orbit. Thus for each atom there are a number of characteristic amounts of energy which correspond to the different possible transi-

DRAMATIS PERSONAE

tions from one orbit to another. These amounts (and no others) can be absorbed; or if the atom has already been excited these amounts can be emitted in the course of returning to the normal state. Here the most characteristic rule of quantum theory comes in. When an atom tips out a lump of energy into the aether the energy always moulds itself into a quantum; that is to say, the energy takes the form of a periodic oscillation or wave such that the amount of the energy divided by the number of oscillations per second is equal to Planck's constant $6 \cdot 55 \cdot 10^{-27}$ erg seconds.

If you wished to determine the pitch of a bell it would be idle to investigate the *quantity* of energy given out. The two measures have no connection, since the same note may be struck loudly or softly. But things are different in the mechanics of an atom, and the amount of energy emitted fixes the pitch or frequency of the resulting radiation. Similarly if light is falling on an atom, its frequency determines the amount of energy offered to the atom for absorption. Only if the amount coincides with one of the possible energies of transition from one orbit to another will the atom accept it. Accordingly the series of characteristic transition energies of the atom corresponds to a series of characteristic frequencies of its radiation. When the radiation is examined with a spectroscope and the different frequencies are thereby laid out side by side for examination, these characteristic frequencies are displayed as the lines of the spectrum.

In general the absorption and emission of visual or ultra-violet light depends on jumps of the valency electrons in the outermost of the occupied orbits. The absorption and emission of X rays depends on jumps from and to one of the innermost orbits, i.e. in the K or L groups.

There is another kind of absorption of radiation, called photo-electric absorption, in which the electron instead of jumping to a higher orbit leaves the atom altogether. This naturally requires more energy than the highest orbit-jump, and the light must correspondingly be of higher frequency. But, provided that it exceeds a certain minimum, no precise amount of energy is required; the electron can carry away any surplus as kinetic energy of its motion. Absorption of this kind accordingly leads to a continuous spectrum which begins just about where the line spectrum leaves off. There is a corresponding emission of radiation when free electrons are captured by ions.

The quanta of radiation which are tipped out by the atoms into the aether in emission, or gathered in by the atoms in absorption, are now generally called photons. How far they can be said to preserve individual existence between their emission by one atom and their absorption by another atom is a very obscure question. But at any rate in emission and absorption each photon behaves as an indivisible atom of radiant energy. Since the amount of energy constituting a photon is proportional to the frequency, we must use high-frequency radiation (X rays or γ rays) if we want a highly concentrated packet of radiant energy to let loose anywhere, e.g. inside an atom.

IV

As far as and beyond the remotest stars the world is filled with aether. It permeates the interstices of the atoms. Aether is everywhere.

How dense is the aether? Is it fluid like water or rigid like steel? How fast is our earth moving through it? Which way do the particles of aether oscillate when an electromagnetic wave travels across it? At one time these were regarded as among the most urgent questions in physics; but at the end of a century's study we have found no answer to any of them. We are, however, convinced that the unanswerableness of these questions is to be reckoned not as ignorance but as knowledge. What we have found out is that aether is not the sort of thing to which such questions would apply. Aether is not a kind of matter. Questions like these could be asked about matter but they could not be asked about *time*, for example; and we must reckon aether as one of the entities to which they are inappropriate.

Since aether is not material it has not any of the usual characteristics of matter—mass, rigidity, etc.—but it has quite definite properties of its own. We describe the state of the aether by symbols, and its characteristic properties by the mathematical equations that the symbols obey.

There is no space without aether, and no aether which does not occupy space. Some distinguished physicists maintain that modern theories no longer require an aether—that the aether has been abolished. I think all they mean is that, since we never have to do with space and aether separately, we can make one word serve for both; and the word they prefer is "space". I suppose they consider that the word aether is still liable to convey the idea of something material. But equally the word space is liable to convey the idea of complete negation. At all events they agree with us in employing an army of mathematical symbols to describe what is going on at any point where the aether is—or, according to them, isn't. "Wheresoever the carcase is, there will the eagles be gathered together", and where the symbols of the mathematical physicist flock, there presumably is some prey for them to settle on, which the plain man at least will prefer to call by a name suggestive of something more than passive emptiness.

Those to whom the word space conveys the idea of characterless void are probably more numerous than those to whom the word aether conveys the idea of a material jelly; so that aether would seem to be the less objectionable term. But it is possible to compromise by using the term "field". The field includes both an electromagnetic field and a gravitational or metrical field; and the army of symbols to which

I have alluded describes these two fields. Space (in its ordinary physical meaning) is the same thing as the metrical field; for the symbols describing the metrical field specify the one characteristic that we are accustomed to ascribe to a space, viz. its geometry (Euclidean or non-Euclidean). In specifying the geometry they specify also the field of gravitation, as Einstein showed in his famous theory. We recognise that there is an inner unity of the electromagnetic and the metrical (gravitational) fields; and the mode of bifurcation of the single unified field into these two component fields is, I think, fairly well understood.*

The change in our conception of the world wrought by the aether or field theory may be illustrated by an incident not infrequent in astronomical observatories. A visitor is handed a photograph of some interesting celestial object. He is puzzled; he turns it this way and that; but he cannot get the hang of the thing. At last the astronomer sees what is the trouble—"I should have explained. This is a *negative*. The dark markings constitute the object; the bright part is only background". The visitor mentally turns the picture inside out, and immediately it makes sense. Something like a turning inside out of our familiar picture of the world is what the aether theory really stands for. Early electrical theories focused attention on an electric fluid flowing along a wire and treated the space outside the wire as mere background. Faraday taught us that, if we would understand the phenomena of electricity, the supposed background—the field outside the wire—was the place to attend to. If you can make this reversal of the picture, turning space from a negative into a positive, so that it is no longer a mere background against which the extension and the motion of matter is perceived but is as much a performer in the world drama as the matter is—then you have the gist of the aether theory whether you use the word "aether" or not.

* *The Nature of the Physical World*, p. 236.

The reversal of the picture is liable to be carried too far. After the great development of the field theory of electromagnetism by Faraday and Maxwell, attention was brought back to the more material aspect by the discovery of the electron and the development of electron theory by Lorentz and Larmor. This reaction in its turn has probably proceeded too far, and it would be a gain if the field aspect were more emphasised. But by gradually diminishing oscillations we are drawing nearer to a unified field-matter theory in which neither the field nor the matter is mere background, and one is seen to be the necessary complement of the other.

V

Hitherto I have not touched the deepest level of ideas in physics. Behind the pictures and models which I have been describing there is a more profound conception of the phenomena, in which the electrons and protons are replaced by waves. This new form of quantum theory originated in a remarkable paper by W. Heisenberg in 1925; the wave conception embodied in it is due more especially to L. de Broglie and E. Schrödinger. It is usually called Wave Mechanics; but the general term quantum theory must be understood to include the new development.

Let us first understand the relation between the particle in the old theory and the wave in the new theory. We have seen that the electron (as a particle) has no size; the conception of size does not apply to it. From a geometrical point of view it is a point, whose sole characteristic is position. But it has also mechanical characteristics, namely momentum and energy (or mass) and a more recondite property called "spin". For our purpose it is sufficient to consider position, since precisely the same ideas apply to the other characteristics.

Regarding then position as the sole characteristic, there is nothing that we can say about the electron unless we know its position. But we may *partially* know its position; we may know that it is in one or other of two places; or we may know that (owing to the attractions and repulsions) it is more likely to be near a proton than near another electron. To describe this partial knowledge, let us imagine a fog whose density at any place is proportional to the *probability* that the electron is at that place. The mass of fog in any volume then represents the probability that the electron is in that volume. The fog extends to every corner of the universe where (according to our knowledge) there is any possibility that the electron may be lurking. If we happen to have exact knowledge of the position we can represent it in the same picture; the fog is then cleared away from all other parts of space and concentrated into a single drop in that position.

We may identify this "drop" with the electron, that is to say it is our pictorial representation of the electron. For we have given the name electron to the entity which occupies the position, and according to our picture the drop is the occupant. But when we go back to partial knowledge a distinction appears. In the new picture the drop diffuses into fog. In the old picture the electron or drop remains concentrated, but we do not quite know where to represent it.

Let us, however, continue to study the fog. In the course of time the position of the electron changes, and equally the positions where it is likely to be are changing. That is to say, the distribution of the fog changes. In an actual medium changes of density are propagated by waves. That is how we come to be concerned with "wave mechanics". Wave mechanics examines the laws of propagation of waves through our fog, and enables us to calculate how in consequence the density of the fog changes in different places. We can thus trace from time to time where the densest part of the fog will be situated. You will remember that the densest part represents the place where the electron is most likely to be found. Thus wave mechanics achieves essentially the same end as ordinary dynamics which traces the motion of the electron as a particle; only it does so in a way adapted to *partial knowledge*. It is useful when our data are given in the form of probabilities or (what comes to the same thing) averages.

The waves of fog must not be confused with the electromagnetic or aether waves which constitute radiation. They are of an altogether different nature.

If you have followed me thus far, you will perhaps say that I have not really reached a profounder level of ideas. I have described an alternative method of treating the problem of the movements of electrons, which has turned out to be the more powerful method in practice; but I have not introduced any real change of conception. As for the fog—if our knowledge is only partial it is natural that our picture should be foggy. I agree. The real change of conception has yet to be introduced.

The crucial point is this. We have discovered the laws of propagation of waves in the fog; we have not discovered the laws of motion of the electron as a particle. Therefore, whatever be the ultimate truth of things, it is the waves not the particles that constitute the world with which the physicist of to-day is dealing.

The older quantum theory which treated the electron as a particle succeeded up to a certain point. But it never got so far as to formulate a system of laws of motion which would cover the jumps of the electron from one orbit to another. It was a collection of strange empirical rules rather than a systematic theory. No one could foresee what would be the next step in its development—what new rule would have to be added. Wave mechanics is a much more unified theory. All its developments proceed naturally from the wave conception, and we do not have to invent *ad hoc* rules as we go along. It is, however, not its aesthetic advantage but its practical success that has led to its universal adoption.

It succeeds better because it attempts less. It does not pretend to tell us where the electron is going next; but it does claim to tell us as much about its future position as is actually involved in the recurrencies of sensory experience. Errors and omissions excepted, wave mechanics enables us to predict sensory experience so far as we have any reason to suppose that sensory experience is predictable; but it does not predict more about the future of the external world than is necessary for this special purpose.

How should we now describe the physical universe or "the universe as it is conceived in modern physics"? It is difficult to speak consistently. I suppose that we ought to mean that conception or formulation which has been generally adopted as giving the most complete agreement with observation. The formulation assigns certain contents and laws, and we are satisfied that by tracing mathematically the consequences of these laws we reproduce the diversity of phenomena, or more strictly the recurrencies of experience, which it is the purpose of physics to analyse; or at least we consider that the failures are not such as to cause uneasiness, bearing in mind that development of the theory is continually proceeding. With that understanding, it cannot be said that the content of the universe as it is conceived in modern physics consists of a number of particles called protons and electrons together with waves of radiation. It is no use assigning contents without laws governing them; and we have not succeeded in formulating a system of law on this basis. In the formulation which must have the credit for the most far-reaching success in scientific prediction, the content of the universe is the "fog", and the basal laws of physics are the laws of propagation of waves of fog—the wave equations. Now that it has become the actual stuff of the universe as it is conceived in physics it is awkward to have to refer to it colloquially as fog. I shall sometimes call it "ψ", that being the symbol by which it enters into our equations, though properly speaking ψ is a measure of the fog rather than the stuff itself. More often I shall call it "probability", i.e. probability of a particle being present; but that implies that there is a universe of particles hovering in the background of our thoughts, although we have seen that it cannot properly be described as the universe conceived in modern physics.

We ought therefore to say that on the present view the content of the universe consists, not of particles, but of waves of ψ. But at the same time it must be realised that a universe composed of ψ waves necessarily contains a large subjective element. Its constituents collect into drops or dissolve into fog according as our knowledge of them happens to be precise or partial. It is a stage whereon the spirit-actors materialise and dissolve as we turn our attention one way and another. There is a provision (Heisenberg's Uncertainty Principle, p. 97) that as the geometrical characteristics of a constituent condense its mechanical characteristics dissolve, so that the actor never comes wholly into focus at one and the same time.

We must concede therefore that "the universe as it is conceived in modern physics" is not identical with what a philosopher would call "the objective physical universe". When we come to think of it there is no reason why it should be. The task of physical science is to disclose the scheme of the recurrencies in the combined experience of conscious beings. We have seen that the universe which constitutes the solution of this problem must necessarily have the characteristics of regularity and externality; we said nothing about objectivity. And so it happens that the aim of science and the search for an objective universe follow the same road up to a certain point and then part company. The scientist then has no choice as to which route to follow; he can only solve the problem for which our experience provides the data.

Thus in saying that wave mechanics corresponds to a profounder level of conception I do not mean that it takes us closer to the objective world behind the phenomena; I mean that it reveals more fully the source of the regularities in our experience, which are conditioned as much by our mode of acquaintance with the objective world as by the constitution of that world. Six years ago* I described wave mechanics as "not a physical theory but a dodge—and a very good dodge too". If I have changed my view at all, it is in regard to the aim of physical theory. If it is still held that the aim of physical theory is to describe objective reality, wave mechanics is not a physical theory in that sense.

The nearest we have got to objective reality is the world of protons and electrons; that is to say, such a world corresponds to the level of conception which physics had reached before it was forced to deviate towards a different aim. Between the universe of our experience and the universe of objective reality probability interposes like a smoke screen.

I will give an example to show that for some purposes an atom constructed out of fog (or ψ) is a *more practical* conception than an atom constructed out of particles. We know that the light waves emitted by an atom have a periodicity which is characteristic of the atom. It is natural to suppose that this periodicity exists within the atom itself and that something concerned in the structure of the atom is oscillating with that period. In the atom constructed of particles (the Bohr model) there is no trace of the period; there is no condition or configuration which goes through a cycle in the period of the emitted light waves. But in the atom constructed out of ψ the period plainly appears; it is the period of the "beats" formed by two sets of ψ waves. I do not want to overstress the significance of this. I mention it as showing that even from a commonsense point of view the change of conception is not wholly a change for the worse. As Heisen-

* *The Nature of the Physical World*, p. 219.

berg pointed out, we have to infer the nature of the inside of an atom from what we observe coming out of it; and since the most definite things coming out of it are certain periodicities shown by the spectral lines, the most logical inference would seem to be that, whatever else there may be in the atom, these periodicities are certainly there.*

VI

I said earlier that the aether (field, space) has no mass; but it would seem that according to a deeper level of conception this is not strictly true. By the general relativity theory mass, momentum and stress are identified with certain components of curvature of space-time—or of the metrical field. Now in a region where there is no recognised matter or electromagnetic field there is still a certain small natural curvature, viz. that specified by the famous "cosmical constant". The mass, momentum and stress equivalent to this curvature ought therefore to be ascribed to whatever we suppose to occupy such a region, i.e. to the space, field or aether—whichever term we are using.

It seems convenient to revive the term aether to express the fact that we do not in any region have to deal with strictly zero density. This turns out to be a crucial consideration in connecting relativity theory with quantum theory. For the operations of quantum theory (wave mechanics) are *multiplicative*. The theory deals with probabilities which are combined by multiplication, not by addition. Now zero is a very awkward number to deal with in multiplicative operations, and similarly empty space is a very awkward sort of abstraction to introduce into quantum theory. The existence everywhere of a residual density provided by the

* I have no high opinion of this argument (for after all the Bohr model did not put anything into the atom that had not been observed coming out of it); but it should appeal to those who stress what are called "commonsense ideas".

natural cosmical curvature thus fits the universe to be the field of application of quantum theory.

We commonly regard completely empty space (devoid of mass and of even the most infinitesimal probability of containing mass) as being the framework common to both theories. Into this empty framework each theory then puts its own characteristic entities; the quantum theory inserts a probability distribution of electrons and protons, and the relativity theory inserts its macroscopically averaged energy tensor of matter and electromagnetic fields. But actually the conception of an empty framework is foreign to both theories, and can indeed only be introduced as a limit. When we examine the standard framework which the theories use—not that which it is commonly imagined they ought to use—the connection leaps to the eye. In relativity theory the norm is, not zero density, but the density corresponding to the natural cosmical curvature. In quantum theory the norm is, not a region certainly devoid of particles, but one in which there is a uniform and isotropic "a priori probability distribution" of the particles and their momenta. The connection of the two theories lies in the identification of these two norms. The mass momentum and stress of the a priori probability distribution in quantum theory is the mass momentum and stress represented by the natural cosmical curvature in relativity theory. We shall deduce important consequences from this later.

Whitehead once said "You cannot have first space and then things to put into it, any more than you can have first a grin and then a Cheshire cat to fit on to it". To adapt the simile to the present state of physics we should have to modify it slightly; we should admit the grin provided that there were a (non-zero) probability of a cat to fit on to it. But leaving aside this minor change the essential truth remains. You cannot have space without things or things without space; and the adoption of thingless space (vacuum) as a standard in most of our current physical thought is a definite hindrance to the progress of physics. By this self-contradictory and irrelevant conception, we have in our current physics made an abstract separation of the theory of space (field) from the theory of things (matter); and now those who are seeking a unified field-matter theory are finding it difficult to join them up again. As I have indicated above the remedy is to use a norm or standard (common to both theories) which does not correspond to complete absence of matter.

"Nature abhors a vacuum." I think that theoretical physics would be wise to follow her example.

CHAPTER III

THE END OF THE WORLD

> Far better 'tis, to die
> the death that flashes gladness,
> than alone, in frigid dignity,
> to live on high.
> Better, in burning sacrifice,
> be thrown against the world
> to perish, than the sky
> to circle endlessly
> a barren stone. *The Shooting Star.**

I

THE title of this chapter is ambiguous. It promises a discussion of the end of the world, but it does not say *which end*. The world—or space-time—is four-dimensional and consequently offers a choice of directions in which we might proceed to look for an end; and it is by no means easy to describe from a purely physical standpoint the direction in which I intend to take you. We shall in fact have to devote most of our attention to this preliminary question "Which end?"

We no longer look for an end of the world in its space dimensions. There is reason to believe that, so far as space dimensions are concerned, the world is of closed spherical type. If we proceed in any direction in space we do not come to an end of space, nor do we continue on and on to infinity; but after travelling a distance, great but not immeasurably great, we find ourselves back at our starting point having "gone round the world". A space that has this re-entrant property is said to be *finite but unbounded*. The surface of a

* Quoted in *Nature*, Aug. 26, 1933. Author unknown.

THE END OF THE WORLD

sphere is an example of a finite but unbounded two-dimensional space; we have to imagine in the universe the same kind of connectivity but with one more dimension. I suppose that even if we can to some extent picture such a bubble space—length, breadth and thickness all lying in the film of the bubble—it is hard to convince ourselves that the picture is not nonsensical as a representation of the space of actual experience. But let me remind you that the familiar idea of space is the idea of the story teller who lives inside our minds. He has never been outside his own doors; he cannot run along to the ends of the nerves and roam into the external world to see what it really is that is arousing our sensory perceptions. When I say that a finite and unbounded type of space is not contradictory to experience, I mean that it is not incompatible with the extremities of our nerves being stimulated by external phenomena in the way requisite to induce the actual sequence of our perceptions. So if finite but unbounded space offers the most satisfactory solution of the cryptogram, there is no reason why we should not accept it—as the solution of the cryptogram.

Spherical space will occupy us in Chapter X and we shall not linger over it here. Let us turn to time. The world is closed in its three space dimensions, but it is open at both ends in its one time dimension. Proceeding from "here" in any spatial direction we ultimately return to "here"; but proceeding from "now" towards the future or the past we never return to "now". There is no bending round of time to bring us back to the moment we set out from. In mathematics we find it convenient to provide for this difference between the closed character of space and the open character of time by means of the symbol $\sqrt{-1}$; those familiar with analytical geometry will recall that the same symbol crops up in differentiating between a closed ellipse and an open hyperbola.

If then we are seeking an end of the world—or an infinite

continuation for all eternity—we must proceed in one of the two time directions. How shall we decide which of the two directions to take? Imagine yourself in some unfamiliar surroundings in space-time—billions of miles from here, billions of years from now—undergoing experiences that you have never undergone before. How would you know which were the earlier and which the later events in those experiences? It is said that in a fog an airman sometimes flies upside down without knowing it. Could one similarly become inverted in time if none of the accustomed indications were discernible? Or is there everywhere and everywhen in the physical universe a signpost with one arm marked "To the Future" and the other arm "To the Past"? My first business is to hunt for this signpost; for if I mistake the way, I shall lead you to what is no doubt an end of the world, but it will be that end which is more usually called "the beginning".

For ordinary purposes the signpost is detected by consciousness. Some would perhaps say that consciousness does not bother about signposts, but wherever it finds itself it hurries off on urgent business in some direction, and the physicist meekly follows its lead and labels the course that it takes "To the Future". It is an important question whether consciousness in selecting its direction is guided by anything in the physical world. If it is so guided we ought to be able to find the particular feature of the physical world which makes it a one-way street for conscious beings. As scientists we are anxious to make the scheme of the physical universe as self-contained as possible. We do not want to be dependent on consciousness, which is outside the scope of physics, for so fundamental a physical distinction as that between past and future. If there were nothing apart from our consciousness that could discriminate future from past we should have to regard the distinction as merely subjective.

Two rather different questions are involved. Anticipating

a little, I may say that a signpost for time *has* been found in the physical universe, so that we are not wholly dependent on the intuition of consciousness. To that extent the distinction of past and future is objective. But our consciousness also insists that the distinction is of a particular kind; it has a kind of dynamic quality which we can feel though we cannot define it. We cannot describe this quality in mathematical symbols, and we cannot therefore expect the physicist to discover it in the external world. Nevertheless the dynamical nature of time—the conception of "becoming"—is so essential a part of our outlook on experience that the purely physical criterion for distinguishing past and future always seems a very inadequate substitute for the going on of time which we perceive in our consciousness. The statement that things "become" from past to future seems to convey a great deal more than the statement that there exists a way of distinguishing past from future.

The view is sometimes held that the dynamic quality of time does not exist in the physical universe and is a wholly subjective impression. Experience presents the physical world as a cinematograph film which is being unrolled in a certain direction; but it is suggested that that is a property of the way the film is inserted into the cinematograph lantern of consciousness, and that there is in the film itself nothing to decide which way it should be unrolled.* If this view were right the "going on of time" ought not to appear in our picture of the external world. Just as we have dropped the old geocentric outlook on the universe, treating it as an idiosyncrasy of our own situation as observers, so we should drop the dynamic presentation of events—the becomingness of things—treating it as a peculiarity of the process of apprehending the world in consciousness. In that case, however, we must be careful not to treat the usual past-to-future

* The two ends are marked distinctively (as we have stated above), but that still leaves open the question which is the right mark to begin at.

presentation of the history of the physical universe as truer or more significant than a future-to-past presentation. In particular we must drop the theory of evolution, or at least set alongside it a theory of anti-evolution as equally true and equally significant.

If anyone holds this view I cannot answer him by argument; I can only cast aspersions on his character. If he is a professional scientist I say to him: "You are a teacher and leader whose duty it is to inculcate a true and balanced outlook. But you teach, or without protest allow your colleagues to teach, a one-sided doctrine of evolution. You teach it, not as a colourless schedule of events, but as if there were something significant, perhaps even morally inspiring, in the development out of formless chaos of the richness and adaptation of our present surroundings. Why do you suppress all reference to the sequence from future to past, which according to you is an equally significant sequence to follow? Why do you not tell us the story of anti-evolution? Show us how from the diverse species existing to-day Nature anti-evolved clumsier forms, more and more unfitted to survive, till she reached the crudity of paleozoic life. Show us how from the system of the stars or the planets Nature anti-evolved chaotic nebulae. Narrate the whole story of anti-progress from future to past, and depict the activity of Nature as a force which takes this great work of architecture around us and—makes a hash of it".

II

Setting aside the guidance of consciousness, we discover a signpost for time in the physical world itself. The signpost is a rather peculiar one, and I would not venture to say that the discovery of the signpost amounts to the same thing as the discovery of an objective "going on of time" in the universe. But at any rate it provides a unique criterion for discriminating between past and future, whereas there is no corresponding absolute distinction between right and left. The signpost depends on a certain measurable physical quantity called entropy. Take an isolated system and measure its entropy at two instants t_1 and t_2; the rule is that the instant which corresponds to the greater entropy is the later. We can thus find out by purely physical measurements whether t_1 is before or after t_2 without trusting to the intuitive perception of the direction of progress of time in our consciousness. In mathematical form the rule is that the entropy S fulfils the law

dS/dt is always positive.

This is the famous Second Law of Thermodynamics.

Entropy may most conveniently be described as a measure of the disorganisation of a system. I do not intend that to be taken as a definition, because disorganisation is a flexible term depending to some extent on our point of view; but in all those processes which increase the entropy of a system we can see chance creeping in where formerly it was excluded, so that conditions which were specialised or systematised become chaotic. Many examples can be given of natural processes which break up an organised system into a random distribution. Plane waves of sunlight all travelling in one direction fall on a white sheet of paper and are scattered in all directions. The direction of the waves becomes disorganised; accordingly there is an increase of entropy. When a solid body moves as a whole, its molecules travel forward together; when it is stopped by hitting something, the molecules begin to move in all directions indiscriminately. It is as though the disciplined march of a regiment suddenly stopped, and it became a jostling throng of individuals all trying to go in different directions. This random motion of the molecules is identified with the heat-energy of the body. Quantitatively the heat produced by impact is the exact equivalent of the lost energy of motion of the body as a whole, but it has a less organised form. Nature keeps strict account of all these little wastages of organisation which are continually occurring; each is debited against the total stock of organisation contained in the universe. The balance is always growing less. One day it will all be used up.

Heat, when concentrated, is not *fully* disorganised energy. A further decrease of organisation occurs when the heat diffuses evenly so as to bring the body and its surroundings to a uniform temperature. In other words heat-energy suffers loss of organisation when it flows from a hotter body to a colder body. This is one of the most common occasions of increase of entropy (disorganisation), for unless the temperature is everywhere uniform heat is always leaking from hotter to colder regions. The fact that a certain amount of organisation is retained in a concentrated store of heat enables us partially to convert heat into visible motion—the reverse of what happens at impact. But only partially. To drive a train we must put into the engine more heat-energy than will appear as energy of motion of the train, the extra quantity being needed to make up for its inferior organisation. In that way without any creation of organisation we furnish enough organised energy to the train; the excess energy, which has been drained of organisation as far as practicable, is turned out as waste into the condenser of the engine.

In using entropy as a signpost for time we must be careful to treat a properly *isolated* system. Isolation is necessary because a system can gain organisation by draining it from other contiguous systems. Evolution shows us that more highly organised systems develop as time goes on. This may be partly a question of definition, for it does not follow that organisation from an evolutionary point of view is to be reckoned according to the same measure as organisation from the entropy point of view. But in any case these highly developed systems may obtain their organisation by a process of collection, not by creation. A human being as he grows from past to future becomes more and more highly organised —or so he fondly imagines. At first sight this appears to contradict the signpost law that the later instant corresponds to the greater disorganisation. But to apply the law we must make an isolated system of him. If we prevent him from acquiring organisation from external sources, if we cut off his consumption of food and drink and air, he will ere long come to a state which everyone would recognise as a state of extreme "disorganisation".

It is possible for the disorganisation of a system to become complete. The state then reached is called thermodynamic equilibrium. Entropy can increase no further and, since the second law of thermodynamics forbids a decrease, it remains constant. Our signpost for time then disappears; and, so far as that system is concerned, time ceases to go on. That does not mean that time ceases to exist; it exists and extends just as space exists and extends, but there is no longer any dynamic quality in it. A state of thermodynamic equilibrium is necessarily a state of death, so that no consciousness will be present to provide an alternative indicator of "time's arrow".

There is no other independent signpost for time in the physical world—at least no other local signpost; so that if we discredit or explain away this property of entropy the distinction of past and future disappears altogether. I base this statement on a law which has become universally accepted in atomic physics, which is called "the Principle of Detailed Balancing".*

III

Having found our signpost, let us look around. Ahead there is ever-increasing disorganisation in the universe. Although the sum total of organisation is diminishing, certain local structures exhibit a more and more highly specialised

* *The Nature of the Physical World*, p. 79.

organisation at the expense of the rest; that is the phenomenon of evolution. But ultimately these must be swallowed up by the advancing tide of chance and chaos, and the whole universe will reach the final state in which there is no more organisation to lose. A few years ago we should have said that it would end as a uniform featureless mass in thermodynamic equilibrium; but that does not take into account what we have recently learnt as to the expansion of the universe. The theory of the expanding universe introduces some differences of description but, I think, no essential difference of principle, and it will be convenient to consider it later in this chapter, adhering for the present to the older ideas. When the final heat-death overtakes the universe time will *extend* on and on, presumably to infinity, but there will be no definable sense in which it can be said to *go on*. Consciousness must have disappeared from the physical world before this stage is reached and, dS/dt having vanished, there will remain nothing to point the way of progress of time. This is the end of the world.

Now let us look in the opposite direction towards the past. Following time backwards we find more and more organisation in the world. If we are not stopped earlier, we go back to a time when the matter and energy of the world had the maximum possible organisation. To go back further is impossible. We have come to another end of space-time—an abrupt end—only according to our orientation we call it "the beginning".

I have no philosophical axe to grind in this discussion. I am simply stating the results to which our present fundamental conceptions of physical law lead. I am much more concerned with the question whether the existing scheme of science is built on a foundation firm enough to stand the strain of extrapolation throughout all time and all space, than with prophecies of the ultimate destiny of material things or with arguments for admitting an act of Creation. I find no difficulty in accepting the consequences of the present scientific theory as regards the future—the heat-death of the universe. It may be billions of years hence, but slowly and inexorably the sands are running out. I feel no instinctive shrinking from this conclusion. From a moral standpoint the conception of a cyclic universe, continually running down and continually rejuvenating itself, seems to me wholly retrograde. Must Sisyphus for ever roll his stone up the hill only for it to roll down again every time it approaches the top? That was a description of Hell. If we have any conception of progress as a whole reaching deeper than the physical symbols of the external world, the way must, it would seem, lie in escape from the Wheel of things. It is curious that the doctrine of the running-down of the physical universe is so often looked upon as pessimistic and contrary to the aspirations of religion. Since when has the teaching that "heaven and earth shall pass away" become ecclesiastically unorthodox?

The extrapolation towards the past raises much graver difficulty. Philosophically the notion of an abrupt beginning of the present order of Nature is repugnant to me, as I think it must be to most; and even those who would welcome a proof of the intervention of a Creator will probably consider that a single winding-up at some remote epoch is not really the kind of relation between God and his world that brings satisfaction to the mind. But I see no escape from our dilemma. One cannot say definitely that future developments of science will not provide an escape; but it would seem that the difficulty arises not so much from a fault in the present system of physical law as in the whole relation of the method of analysis of experience employed in physical science to the actualities with which it deals. The dilemma is this: Surveying our surroundings we find them to be far from a "fortuitous concourse of atoms". The picture of the world as drawn in existing physical theories shows an arrangement of the individual atoms and photons which if it originated by a chance coincidence would be excessively improbable. The odds against it are multillions to 1. (I use "multillions" as a general term for a number which, if written out in full in the usual decimal notation, would fill all the books in a large library.) This non-random feature of the world might possibly be identified with purpose or design; let us, however, non-committally call it anti-chance. We are unwilling to admit in physics that there is any anti-chance in the reactions between the billions of atoms and quanta in the inorganic systems that we study; and indeed all our experimental evidence goes to show that these are governed by the laws of chance. Accordingly we do not recognise anti-chance in the laws of physics, but only in the data to which those laws are applied. In the corresponding mathematical treatment we exclude anti-chance from the differential equations of physics and relegate it to the boundary conditions—for it has to be brought in somewhere. One cannot help feeling that this segregation of the chance from the anti-chance is a characteristic rather of our method of attacking the problem than of the objective universe itself. It is as though we ironed out a region large enough to include our more immediate experience at the cost of puckering in the regions outside. We have swept away the anti-chance from the field of our current physical problems, but we have not got rid of it. When some of us are so misguided as to try to get back milliards of years into the past we find the sweepings piled up like a high wall, forming a boundary—a beginning of time—which we cannot climb over.

Without insisting dogmatically on the finality of the second law of thermodynamics, we must emphasise that it is very deeply rooted in physics. The engineer dealing with the practical problems of the heat engine, the quantum physicist discussing the laws of radiation, the astronomer investigating the interior of a star, the student of cosmic rays tracing perhaps the disintegrations of atoms in space beyond the galaxy, have all pinned their faith to the rule that the disorganisation or random element can increase but never diminish. This faith is not unreasonable when we recall that to abandon the second law of thermodynamics means that we uproot the signpost of time.

I have sometimes been taken to task for not sufficiently emphasising in my discussions of these problems that the laws concerning entropy are a matter of probability, not of certainty. I said above that if we observe a system at two instants, the instant corresponding to the greater entropy is the later. Strictly speaking, I ought to have said that (for a smallish system) the chances are, say, 10^{20} to 1 that it is the later. For by a highly improbable coincidence the multitudinous particles might at the later instant accidentally arrange themselves in a distribution with as much organisation as at the earlier instant; just as in shuffling a pack of cards there is a possibility that we may accidentally arrange the cards in suits or sequences. Some critics seem to have been shocked at my lax morality in making the former statement when I was well aware of the 1 in 10^{20} chance of its being wrong. Let me make a confession. I have in the past twenty-five years written a number of scientific papers and books, broadcasting a good many statements about the physical world. I am afraid there are not many of these statements for which I can claim that the chance of being wrong is no more than 1 in 10^{20}. My average risk is more like 1 in 10—or is that too boastful an estimate? Certainly if it turns out that nine-tenths of what I tell you in this book is correct, I am either very fortunate or else very platitudinous. I think that if we were not allowed to make statements which had a 1 in 10^{20} chance of being untrue, conversation would languish somewhat. Presumably the only persons entitled to open their lips would be the pure mathematicians.

IV

One way out of the difficulty of an abrupt beginning has sometimes found favour. I oppose it not through any desire to retain the present dilemma but because I do not think it is a genuine loophole. It depends on the occurrence of chance fluctuations. If we have a number of entities moving about at random, they will in the course of time go through every possible configuration; so that even the most orderly, the most non-chance configuration, will occur by chance if we wait long enough—

> There once was a brainy baboon
> Who always breathed down a bassoon
>   For he said "It appears
>   That in billions of years
> I shall certainly hit on a tune".

When the world has reached complete disorganisation (thermodynamic equilibrium) there is still infinite time ahead of it, and its elements will have the opportunity to take up every possible configuration again and again. If we wait long enough a number of atoms will, just by chance, arrange themselves as the atoms are now arranged in this room; and, just by chance, the same sound waves will come from one of the systems of atoms as are now emerging from my lips; they will strike other systems of atoms arranged, just by chance, to resemble you, and in the same stages of attention or somnolence. This mock delivery of the present course of Messenger Lectures will repeat itself many times over—an infinite number of times in fact—before t reaches ∞. Do not ask me whether I really believe, or expect you to believe, that this will happen—*

> Logic is logic. That's all I say.

* See p. 68.

So after the world has reached thermodynamic equilibrium the entropy remains steady at its maximum value, except that once in a blue moon an absurdly small chance comes off and the entropy drops appreciably below its maximum value. When this fluctuation has died out there will again be a very long wait for another coincidence giving another fluctuation. It will take multillions of years, but we have all eternity before us. There is no limit to the possible amount of the fluctuation; and, if we wait long enough, there will be a fluctuation which will take the universe as far from thermodynamic equilibrium as it is at the present moment.

The suggestion is that we are now on the down-slope of one of these chance fluctuations. Is it an accident that we happen to be running down the slope and not toiling up the slope? Not at all. So far as the physical universe is concerned the direction of time has been *defined* to be that in which disorganisation increases, so that on whichever slope of the mountain we stand the signpost "To the Future" points downhill. In fact, on this theory, the going on of time is not a property of time in general but of the slope of the fluctuation on which we stand.

We can always argue that anything that can be done by arrangement or by specific cause can also be done by chance, provided that it is agreed not to count the failures. In this case the theory postulates a state of things involving an exceedingly rare coincidence, but it also provides an infinite time during which the coincidence might (or, it is suggested, *must*) occur. Nevertheless I feel sure that the argument is fallacious.

If we put a kettle of water on the fire, there is a chance that the water will freeze; for the physical theory of the flow of heat indicates that there is very high probability that heat will flow from the fire to the kettle but also a trifling chance that it will flow the other way. If man goes on putting kettles on the fire long enough the chance will one day come off, and the individual concerned will be surprised to find a lump of ice in his kettle. But it will not happen to *me*. So confident of this am I that even if to-morrow I find ice instead of boiling water in the kettle I shall not explain it that way. Probably I shall exclaim "The devil's in it". That indeed would be a more rational explanation. At present I do not believe that devils interfere with cooking arrangements or other experimental proceedings because I am convinced by experience that Nature obeys certain uniformities which we call laws. I am convinced because these laws have been tested over and over again. But it is possible that every single observation from the beginning of science has just happened to fit in with the laws by a chance coincidence. That would, of course, be a highly improbable coincidence, but I calculate that it is not quite so improbable as the coincidence involved in my kettle of water freezing. So if the event happens and I can think of no other explanation, I shall have to choose between two highly improbable coincidences: (*a*) that there is no foundation for the system of physical law accepted in science, and that the apparent uniformity of Nature observed up to now is merely a coincidence; (*b*) that the accepted laws of Nature are true but that I have happened upon a phenomenon due to an improbable coincidence. Both explanations do great violence to probability, but I think that the former is numerically the less unlikely. You will see that when the adverse chance rises to multillions a new relation arises between what we commonly term the "improbable" and the "impossible". I reckon a sufficiently improbable coincidence occurring within the supposed laws of Nature as more disastrous than an actual violation of the laws; because my whole reason for accepting the laws of Nature rests on the assumption that improbable coincidences do not happen—or at least that they do not happen in my experience. No doubt coincidences described as "extremely improbable" occur to all of us, but the im-

THE END OF THE WORLD

probability is of an utterly different order of magnitude from that concerned in the present discussion.

For that reason if logic assures me that a mock performance of these lectures will occur just by fortuitous arrangement of atoms sometime before $t=\infty$, I would reply that I cannot possibly accept that as an explanation of the performance of the lectures in $t=1934$. We must be a little careful over this, because there is a trap for the unwary. The crude argument is that at a particular epoch (1934) the chance of a fortuitous deviation of entropy from its maximum value sufficient to admit the phenomenon is too small to be considered seriously, and that the fluctuation must therefore be ascribed to anti-chance. But the year 1934 is not a random date between $t=-\infty$ and $t=+\infty$. We must not argue that because fluctuations of the present magnitude occupy only $1/x$th of the time between $t=-\infty$ and $t=+\infty$, therefore the chances are x to 1 against such a fluctuation existing in the year 1934. For our present purpose the important characteristic of the year 1934 is that it is selected as belonging to a period during which there exist in the universe beings capable of speculating about the universe and its fluctuations. It is clear that such creatures could not exist in conditions near thermodynamical equilibrium. Therefore it is perfectly fair for the supporters of this suggestion to wipe out of the calculation all those multillions of years during which the fluctuations are less than the minimum required to permit of the evolution and existence of mathematical physicists. That greatly diminishes x; but the odds are still overpowering. The *crude* assertion would be that (unless we admit something which is not chance in the architecture of the universe) it is practically certain that the universe will be found to be almost in the state of maximum disorganisation. The *amended* assertion is that (unless we admit something which is not chance in the architecture of the universe) it is practically certain that a universe which contains mathematical physicists will be found to be almost in the state of maximum disorganisation which is not inconsistent with the existence of such creatures. I think it is clear that neither the crude nor the amended version applies. It appears necessary therefore to admit anti-chance; and from our present scientific standpoint the best we can do with it is to sweep it up into a heap at the beginning of time, as I have already described.

V

The irreversible dissipation of energy in the universe has been a recognised doctrine of science since 1852 when it was formulated explicitly by Lord Kelvin. Kelvin drew the same conclusions about the beginning and end of things as those given here—except that, since less attention was paid to the universe in those days, he considered the earth and the solar system. The general ideas have not changed much in eighty years; but the recognition of the finitude of space and the recent theory of the expanding universe now involve some supplementary considerations.

The conclusion that the total entropy of the universe at any instant is greater than at a previous instant dates from a time when an "instant" was conceived to be an absolute time-partition extending throughout the universe. We have to reconsider the matter now that Einstein has abolished these absolute instants; but it appears that no change is required. I think I am right in saying that it is not necessary that the instants should be absolute, or that the time t referred to in dS/dt should be a form of absolute time. For the first instant we can choose any arbitrary space-like section of space-time (smooth or crinkled), and for the second instant any other space-like section which does not intersect the first. One of these instants will be later throughout than the other;* and

* That is to say, all observers, whatever their position and motion, will encounter them in the same order.

the total entropy of the universe integrated over the later instant will be greater than over the earlier instant. This generalisation is made possible by the fact that the energy or matter which carries the disorganisation cannot travel from place to place faster than light.

The consequences of introducing the expansion of the universe are more difficult to foresee. Fundamental questions are raised as to the appropriate way of defining entropy when the background conditions are no longer invariable. I believe that the progress of the theory in other directions in the next few years will place us in a better position to treat the thermodynamical problem which it raises, and I prefer not to try to anticipate its conclusions.

Meanwhile it is important to notice that the expansion of the universe is another irreversible process. It is a one-way characteristic like the increase of disorganisation. Just as the entropy of the universe will never return to its present value, so the volume of the universe will never return to its present value. From the expansion of the universe we reach independently the same outlook as to the beginning and end of things that we have here reached by considering the increase of entropy. In particular the conclusion seems almost inescapable that there must have been a definite beginning of the present order of Nature. The theory of the expanding universe adds something new, namely an estimate of the date of this beginning. We shall see in Chapter x that from the scientific point of view it is uncomfortably recent—scarcely more than 10,000 million years ago.

In the expanding universe we can decide which of two instants is the later by the criterion that the later instant corresponds to the larger volume of the universe. (The instants are defined as before to be two non-intersecting space-like sections of space-time.) This provides an alternative signpost for time. But it is only applicable to time taken throughout the universe as a whole. The position of entropy as the unique *local signpost* remains unaffected. The fact that the direction of time for the universe, regarded as a single system, is indicated both by increasing volume and by increasing entropy suggests that there is some undiscovered relation between the two criteria. That is one of the points on which we may expect more light in the next few years.

By accepting the theory of the expanding universe we are relieved of one conclusion which we had felt to be intrinsically absurd. It was argued (p. 62) that every possible configuration of atoms must repeat itself at some distant date. But that was on the assumption that the atoms will have only the same choice of configurations in the future that they have now. In an expanding space any particular congruence becomes more and more improbable. The expansion of the universe creates new possibilities of distribution faster than the atoms can work through them, and there is no longer any likelihood of a particular distribution being repeated. If we continue shuffling a pack of cards we are bound sometime to bring them into their standard order—but not if the conditions are that every morning one more card is added to the pack.

So I think after all there will not be a second (accidental) delivery of these Messenger Lectures this side of eternity.

VI

To what extent are conscious beings subject to the second law of thermodynamics? The way in which conscious purpose might intervene was pointed out by Clerk Maxwell who invented a famous "sorting demon". Two adjacent vessels contain gas at the same uniform temperature; between them there is a very small door. At the door there stands a demon. Whenever he sees in the left-hand vessel an unusually fast-moving molecule approaching the door, he opens it so that the molecule goes through into the right-hand vessel;

for slow-moving molecules he keeps the door shut and they rebound into the left-hand vessel. Similarly he allows slow-moving molecules from the right-hand vessel to pass through into the left. The result is that he concentrates fast motion in the right-hand vessel and slow motion in the left-hand vessel; or since the speed of molecular motion corresponds to temperature, the right-hand vessel becomes hot and the left-hand vessel cool. Ideally he might do this without expending any energy, since the door might be poised so that an infinitesimal effort would open or shut it. But to create a difference of temperature of this kind is a gain of organisation; it is the opposite of the natural process of disorganisation by the flow of heat from a hot to a cold region. Maxwell's demon overrides the second law of thermodynamics.

When in Nature a hot body and a cold body are in contact, we find that, as time goes on, the hot body cools and the cold body becomes warmer until the temperatures are equalised. That is if we have not mistaken the signpost of time. But if we happened to have lost our bearings and were viewing time backwards, we should see the two bodies first at equal temperatures, and then one becoming hotter and the other colder—precisely the effect that Maxwell's demon achieves. Thus effectively the demon reverses the signpost of time. Being a *sorting* agent, he is the embodiment of anti-chance; and in his domain time appears to run the opposite way from that taken in normal systems under the government of chance.

The mind of man, in virtue of its conscious purpose, must play to some extent the part of Maxwell's sorting demon. But we must not forget that mind can only make its purposes effective in the physical world through its association with a body; and whilst the mind may (or may not) be increasing organisation the body is always increasing disorganisation. It is obvious that (reckoned in physical measure) the organisation brought about by our conscious purpose is very small compared with that which we consume in eating and breathing, so that taken as a whole we do not stem the current of increasing disorganisation. One may hazard the suggestion that this is not an accidental limitation, but that even the purposive activity of human beings is subject to the second law of thermodynamics; and that the relation of mind and body is such that of necessity the amount of organisation which the one can put into the world is limited by the amount that the other takes out of the world.

I have sometimes wondered whether it would not be possible to baffle Maxwell's sorting demon by one of the modern developments of atomic physics, viz. Heisenberg's Uncertainty Principle (p. 97). This asserts that a knowledge of exact position of a particle is incompatible with a knowledge of exact velocity. I picture the demon scanning the approaching molecules for those of large velocity. Since for his purpose he has to know their velocities, he must by the foregoing principle be uncertain of their positions. He does not know how far off they are and how soon they will reach the door. So he has to chance the time of opening; and when he opens it for the expected high-speed molecule it is quite likely that a low-speed molecule will slip through. But I am afraid the demon is too clever for me. In some circumstances at any rate, his knowledge of both position and velocity, though inexact, would be sufficient for the purpose of his job; and his mistakes would not be so frequent as to prevent a progressive separation of high and low speed molecules. Apparently the only way of frustrating the demon is to tether him to flesh and blood so that his body spends the anti-chance that his mind produces.

I suppose that to justify the title of this chapter I ought to conclude with a prophecy of what the End of the World will be like. It is, of course, not the purpose of our investigation to make such prophecies. However, after our serious efforts we can perhaps relax. It used to be thought that in the end all the matter of the universe would collect into one rather dense ball at uniform temperature. But the doctrine of spherical space, and more especially the recent results as to the expansion of the universe, have changed all that. There are unsettled points which prevent a definite conclusion; so I will content myself with stating one of several possibilities. It has been widely supposed that the ultimate fate of protons and electrons is to annihilate one another, and release the energy of their constitution in the form of radiation. If so it would seem that the universe will finally become a ball of radiation, becoming more and more rarified and passing into longer and longer wave-lengths. The longest waves of radiation are Hertzian waves of the kind used in broadcasting. About every 1500 million years this ball of radio waves will double its diameter; and it will go on expanding in geometrical progression for ever. Perhaps then I may describe the end of the physical world as—one stupendous broadcast.

CHAPTER IV

THE DECLINE OF DETERMINISM

Thus from the outset we can be quite clear about one very important fact, namely, that the validity of the law of causation for the world of reality is a question that cannot be decided on grounds of abstract reasoning.
MAX PLANCK, *Where is Science Going?* p. 113.

The new theory appears to be well founded on observation, but one may ask whether *in the future*, by development or refinement, it may not be made deterministic again. As to this it must be said: It can be shown by rigorous mathematics that the accepted formal theory of quantum mechanics does not admit of any such extension. If anyone clings to the hope that determinism will ever return, he must hold the existing theory to be false in substance; it must be possible to disprove experimentally definite assertions of this theory. The determinist should therefore not protest but experiment.
MAX BORN, *Naturwissenschaften*, 1929, p. 117.

Whilst the feeling of free-will dominates the life of the spirit, the regularity of sensory phenomena lays down the demand for causality. But in both domains simultaneously the point in question is an idealisation, whose natural limitations can be more closely investigated, and which determine one another in the sense that the feeling of volition and the demand for causality are equally indispensable in the relation between Subject and Object which is the kernel of the problem of perception.
NIELS BOHR, *Naturwissenschaften*, 1930, p. 77.

We must await the further development of science, perhaps for centuries, before we can design a true and detailed picture of the interwoven texture of Matter, Life and Soul. But the old classical determinism of Hobbes and Laplace need not oppress us any longer.
HERMANN WEYL, *The Open World*, p. 55.

I

TEN years ago practically every physicist of repute was, or believed himself to be, a determinist, at any rate so far as inorganic phenomena are concerned. He believed he had come across a scheme of strict causality regulating the sequence of phenomena. It was considered to be the primary aim of science to fit as much of the universe as possible into such a scheme; so that, as a working belief if not as a philosophical conviction, the causal scheme was always held to be applicable in default of overwhelming evidence to the contrary. In fact, the methods, definitions and conceptions of physical science were so much bound up with the hypothesis of strict causality that the limits (if any) of the scheme of causal law were looked upon as the ultimate limits of physical science. No serious doubt was entertained that this determinism covered all inorganic phenomena. How far it applied to living or conscious matter or to consciousness itself was a matter of individual opinion; but there was naturally a reluctance to accept any restriction of an outlook which had proved so successful over a wide domain.

Then rather suddenly determinism faded out of theoretical physics. Its exit has been received in various ways. Some writers are incredulous and cannot be persuaded that determinism has really been eliminated from the present foundations of physical theory. Some think that it is no more than a domestic change in physics, having no reactions on general philosophic thought. Some decide cynically to wait and see if determinism fades in again.

The rejection of determinism is in no sense an abdication of scientific method. It is rather the fruition of a scientific method which had grown up under the shelter of the old causal method and has now been found to have a wider range. It has greatly increased the power and precision of the mathematical theory of observed phenomena. On the other hand I cannot agree with those who belittle the philosophical significance of the change. The withdrawal of physical science from an attitude it had adopted consistently for more than 200 years is not to be treated lightly; and it provokes a reconsideration of our views as to one of the most perplexing problems of our existence.

In a subject which arouses so much controversy it seems well to make clear at the outset certain facts regarding the extent of the change as to which there has frequently been a misunderstanding. Firstly, it is not suggested that determinism has been disproved. What we assert is that physical science is no longer based on determinism. Is it difficult to grasp this distinction? If I were asked whether astronomy has disproved the doctrine that "the moon is made of green cheese" I might have some difficulty in finding really conclusive evidence; but I could say unhesitatingly that the doctrine is not the basis of present-day selenography. Secondly, the denial of determinism, or as it is often called "the law of causality", does not mean that it is denied that effects may proceed from causes. The common regular association of cause and effect is a matter of experience; the law of causality is an extreme generalisation suggested by this experience. Such generalisations are always risky. To suppose that in doubting the generalisation we are denying the experience is like supposing that a person who doubts Newton's (or Einstein's) law of gravitation denies that apples fall to the ground. The first criterion applied to any theory, deterministic or indeterministic, is that it must account for the regularities in our sensory experience—notably our experience that certain effects regularly follow certain causes. Thirdly, the admission of indeterminism in the physical universe does not immediately clear up all the difficulties—not even all the physical difficulties—connected with Free Will. But it so far modifies the problem that the door is not barred and bolted for a solution less repugnant to our deepest intuitions than that which has hitherto seemed to be forced upon us.

Let us be sure that we agree as to what is meant by determinism. I quote three definitions or descriptions for your consideration. The first is by a mathematician (Laplace):

We ought then to regard the present state of the universe as the effect of its antecedent state and the cause of the state that is to follow. An intelligence, who for a given instant should be acquainted with all the forces by which Nature is animated and with the several positions of the entities composing it, if further his intellect were vast enough to submit those data to analysis, would include in one and the same formula the movements of the largest bodies in the universe and those of the lightest atom. Nothing would be uncertain for him; the future as well as the past would be present to his eyes. The human mind in the perfection it has been able to give to astronomy affords a feeble outline of such an intelligence.... All its efforts in the search for truth tend to approximate without limit to the intelligence we have just imagined.

The second is by a philosopher (C. D. Broad):

"Determinism" is the name given to the following doctrine. Let S be any substance, ψ any characteristic, and t any moment. Suppose that S is in fact in the state σ with respect to ψ at t. Then the compound supposition that everything else in the world should have been exactly as it in fact was, and that S should instead have been in one of the other two alternative states with respect to ψ is an impossible one. [The three alternative states (of which σ is one) are: to have the characteristic ψ, not to have it, and to be changing.]

The third is by a poet (Omar Khayyam):

With Earth's first Clay They did the Last Man's knead,
And then of the Last Harvest sow'd the Seed:
Yea, the first Morning of Creation wrote
What the Last Dawn of Reckoning shall read.

I regard the poet's definition as my standard. There is no doubt that his words express what is in our minds when we refer to determinism. In saying that the physical universe as now pictured is not a universe in which "the first morning of creation wrote what the last dawn of reckoning shall read", we make it clear that the abandonment of determinism is no technical quibble but is a fundamental change of outlook. The other two definitions need to be scrutinised suspiciously; we are afraid there may be a catch in them. In fact I think there is a catch in them.*

It is important to notice that all three definitions introduce the time element. Determinism postulates not merely causes but *pre-existing* causes. Determinism means predetermination. Hence in any argument about determinism the dating of the alleged causes is all-important; we must challenge them to produce their birth-certificates.

In the passage quoted from Laplace a definite aim of science is laid down. Its efforts "tend to approximate without limit to the intelligence we have just imagined", i.e. an intelligence who from the present state of the universe could foresee the whole of future progress down to the lightest atom. This aim was accepted without question until recent times. But the practical development of science is not always in a direct line with its ultimate aims; and about the middle of the nineteenth century there arose a branch of physics (thermodynamics) which struck out in a new direction. Whilst striving to perfect a system of law that would predict what *certainly* will happen, physicists also became interested in a system which predicts what *probably* will happen. Alongside the super-intelligence imagined by Laplace for whom "nothing would be uncertain" was placed an intelligence for whom nothing would be certain but some things would be exceedingly probable. If we could say of this latter being that for him *all* the events of the future were known with exceedingly high probability, it would be mere pedantry to distinguish him from Laplace's being who is supposed to know them with certainty. Actually, however, the new being is supposed

---

* The catch that I suspect in Broad's definition is that it seems to convey no meaning without further elucidation of what is meant by the supposition being an *impossible* one. He does not mean impossible because it involves a logical contradiction. The supposition is not rejected as being contrary to logic nor as contrary to fact, but for a third reason undefined.

to have glimpses of the future of varying degrees of probability ranging from practical certainty to entire indefiniteness according to his particular field of study. Generally speaking his predictions never approach certainty unless they refer to an average of a very large number of individual entities. Thus the aim of science to approximate to this latter intelligence is by no means equivalent to Laplace's aim. I shall call the aim defined by Laplace the *primary* aim, and the new aim introduced in the science of thermodynamics the *secondary* aim.

We must realise that the two aims are distinct. The prediction of what will probably occur is not a half-way stage in the prediction of what will certainly occur. We often solve a problem approximately, and subsequently proceed to second and third approximations, perhaps finally reaching an exact solution. But here the probable prediction is an end in itself; it is not an approximate attempt at a certain prediction. The methods differ fundamentally, just as the method of diagnosis of a doctor who tells you that you have just three weeks to live differs from that of a Life Insurance Office which tells you that your expectation of life is 18·7 years. We can, of course, occupy ourselves with the secondary aim without giving up the primary aim as an ultimate goal; but a survey of the present state of progress of the two aims produces a startling revelation.

The formulae given in modern textbooks on quantum theory—which are continually being tested by experiment and used to open out new fields of investigation—are exclusively concerned with probabilities and averages. This is quite explicit. The "unknown quantity" which is chased from formula to formula is a probability or averaging factor. The quantum theory therefore contributes to the secondary aim, but adds nothing to the primary Laplacian aim which is concerned with causal certainty. But further it is now recognised that the classical laws of mechanics and electromagnetism (including the modifications introduced by relativity theory) are simply the limiting form assumed by the formulae of quantum theory when the number of individual quanta or particles concerned is very large. This connection is known as Bohr's Correspondence Principle. The classical laws are not a fresh set of laws, but are a particular adaptation of the quantum laws. So they also arise from the secondary scheme. We have already mentioned that it is when a very large number of individuals are concerned that the predictions of the secondary scheme have a high probability approaching certainty. Consequently the domain of the classical laws is just that part of the whole domain of secondary law in which the probability is so high as to be practically equivalent to certainty. That is how they came to be mistaken for causal laws whose operation is definitely certain. Now that their statistical character is recognised they are lost to the primary scheme. When Laplace put forward his ideal of a completely deterministic scheme he thought he already had the nucleus of such a scheme in the laws of mechanics and astronomy. That nucleus has now been transferred to the secondary scheme. Nothing is left of the old scheme of causal law, and we have not yet found the beginnings of a new one.

Measured by advance towards the secondary aim, the progress of science has been amazingly rapid. Measured by advance towards Laplace's aim its progress is just *nil*.

Laplace's aim has lapsed into the position of other former aims of science—the discovery of the elixir of life, the philosopher's stone, the North-West Passage—aims which were a fruitful inspiration in their time. We are like navigators on whom at last it has dawned that there are other enterprises worth pursuing besides finding the North-West Passage. I need hardly say that there are some old mariners who regard these new enterprises as a temporary diversion and predict an early return to the "true aim of geographical exploration".

II

Let us examine how the new aim of physics originated. We observe certain regularities in the course of phenomena and formulate these as laws of Nature. Laws can be stated positively or negatively, "Thou shalt" or "Thou shalt not". For the present purpose we shall formulate them negatively. Here are two regularities in the sensory experience of most of us:

(a) We never come across equilateral triangles whose angles are unequal.

(b) We never come across thirteen hearts in a hand dealt to us at Bridge.

In our ordinary outlook we explain these regularities in fundamentally different ways. We say that the first holds because a contrary experience is *impossible*; the second because a contrary experience is *too improbable*.

This distinction is theoretical. There is nothing in the observations themselves to suggest to which type a particular observed regularity belongs. We recognise that "impossible" and "too improbable" are both adequate explanations of any observed uniformity of experience; and formerly physics rather haphazardly explained some uniformities one way and others the other way. But now the whole of physical law (so far discovered) is found to be comprised in the secondary scheme which deals only with probabilities; and the only reason assigned for any regularity is that the contrary is too improbable. Our failure to find equilateral triangles with unequal angles is because such triangles are too improbable. Of course, I am not here referring to the theorem of pure geometry; I am speaking of a regularity of sensory experience and refer therefore to whatever measurement is supposed to confirm this property of equilateral triangles as being true of actual experience. Our measurements regularly confirm it to the highest accuracy attainable, and no doubt will always do so; but according to the present physical theory that is because a failure could only occur as the result of an extremely unlikely coincidence in the behaviour of the vast number of particles concerned in the apparatus of measurement.

The older view, as I have said, recognised two types of natural law. The earth keeps revolving round the sun because it is impossible that it should run away. That is the primary or deterministic type. Heat flows from a hot body to a cold body because it is too improbable that it should flow the other way. That is the secondary or statistical type. On the modern theory both regularities belong to the statistical type—it is too improbable that the earth should run away from the sun.*

So long as the aim of physics is to bring to light a deterministic scheme, the pursuit of secondary law is a blind alley since it leads only to probabilities. The determinist is not content with a law which ordains that, given reasonable luck, the fire will warm me; he agrees that that is the probable event, but adds that somewhere at the base of physics there are other laws which ordain just what the fire will do to me, luck or no luck.

To borrow an analogy from genetics, determinism is a *dominant character*. Granting a system of primary law, we can (and indeed must) have secondary indeterministic laws derivable from it stating what will probably happen under that system. So for a long time determinism watched with equanimity the development within itself of a subsidiary indeterministic system of law. What matter? Deterministic law remains dominant. It was not foreseen that the child would grow to supplant its parent. There is a game called "Think of a number". After doubling, adding, and other

* "Impossible" therefore disappears from our vocabulary except in the sense of involving a logical contradiction. But the logical contradiction or impossibility is in the description, not in the phenomenon which it attempts but (on account of the contradiction) fails to describe.

calculations, there comes the direction "Take away the number you first thought of". Determinism is now in the position of the number we first thought of.

The growth of secondary law whilst still under the dominant deterministic scheme was remarkable, and whole sections of physics were transferred to it. There came a time when in the most progressive branches of physics it was used exclusively. The physicist might continue to profess allegiance to primary law but he ceased to use it. Primary law was the gold stored in the vaults; secondary law was the paper currency actually used. But everyone still adhered to the traditional view that paper currency needs to be backed by gold. As physics progressed the occasions when the gold was actually produced became rarer until they ceased altogether. Then it occurred to some of us to question whether there still was a hoard of gold in the vaults or whether its existence was a mythical tradition. The dramatic ending of the story would be that the vaults were opened and found to be empty. The actual ending is not quite so simple. It turns out that the key has been lost, and no one can say for certain whether there is any gold in the vaults or not. But I think it is clear that, with either termination, present-day physics is *off the gold standard.*

III

The nature of the indeterminism now admitted in the physical world will be considered in more detail in the next chapter. I will here content myself with an example showing its order of magnitude. Laplace's ideal intelligence could foresee the future positions of objects from the heaviest bodies to the lightest atoms. Let us then consider the lightest particle we know, viz. the electron. Suppose that an electron is given a clear course (so that it is not deflected by any unforeseen collisions) and that we know all that can be known about it at the present instant. How closely can we foretell its position one second later? The answer is that (in the most favourable circumstances) we can predict its position to within about $1\frac{1}{2}$ inches—not closer. That is the nearest we can approximate to Laplace's super-intelligence. The error is not large if we recall that during the second covered by our prediction the electron may have travelled 10,000 miles or more.

The uncertainty would, however, be serious if we had to calculate whether the electron would hit or miss a small target such as an atomic nucleus. To quote Prof. Born: "If Gessler had ordered William Tell to shoot a hydrogen atom off his son's head by means of an α particle and had given him the best laboratory instruments in the world instead of a cross-bow, Tell's skill would have availed him nothing. Hit or miss would have been a matter of chance".

For contrast take a mass of ·001 milligram—which must be nearly the smallest mass handled macroscopically. The indeterminacy is much smaller because the mass is larger. Under similar conditions we could predict the position of this mass a thousand years hence to within $\frac{1}{5000}$ of a millimetre.

This indicates how the indeterminism which affects the minutest constituents of matter becomes insignificant in ordinary mechanical problems, although there is no change in the basis of the laws. It may not at first be apparent that the indeterminacy of $1\frac{1}{2}$ inches in the position of the electron after the lapse of a second is of any great practical importance either. It would not often be important for an electron pursuing a straight course through empty space; but the same indeterminism occurs whatever the electron is doing. If it is pursuing an orbit in an atom, long before the second has expired the indeterminacy amounts to atomic dimensions; that is to say, we have altogether lost track of the electron's position in the atom. Anything which depends on the relative location of electrons in an atom is unpredictable more than a minute fraction of a second ahead.

For this reason the break-down of an atomic nucleus, such as occurs in radio-activity, is not predetermined by anything in the existing scheme of physics. All that the most complete theory can prescribe is how frequently configurations favouring an explosion will occur on the average; the individual occurrences of such a configuration are unpredictable. In the solar system we can predict fairly accurately how many eclipses of the sun (i.e. how many recurrences of a special configuration of the earth, sun and moon) will happen in a thousand years; or we can predict fairly accurately the date and time of each particular eclipse. The theory of the second type of prediction is not an elaboration of the theory of the first; the occurrence of individual eclipses depends on celestial mechanics, whereas the frequency of eclipses is purely a problem of geometry. In the atom, which we have compared (p. 29) to a miniature solar system, there is nothing corresponding to celestial mechanics—or rather mechanics is stifled at birth by the magnitude of the indeterminacy—but the geometrical theory of frequency of configurations remains analogous.

The future is never entirely determined by the past, nor is it ever entirely detached. We have referred to several phenomena in which the future is *practically determined*; the break-down of a radium nucleus is an example of a phenomenon in which the future is *practically detached* from the past.

But, you will say, the fact that physics assigns no characteristic to the radium nucleus predetermining the date at which it will break up, only means that that characteristic has not yet been discovered. You readily agree that we cannot predict the future in all cases; but why blame Nature rather than our own ignorance? If the radium atom were an exception, it would be natural to suppose that there is a determining characteristic which, when it is found, will bring it into line with other phenomena. But the radio-active atom was not brought forward as an exception; I have mentioned it as an extreme example of that which applies in greater or lesser degree to all kinds of phenomena. There is a difference between explaining away an exception and explaining away a rule.

The persistent critic continues, "You are evading the point. I contend that there are characteristics unknown to you which completely predetermine not only the time of break-up of the radio-active atom but all physical phenomena. How do you know that there are not? You are not omniscient?" I can understand the casual reader raising this question; but when a man of scientific training asks it, he wants shaking up and waking. Let us try the effect of a story.

About the year 2000, the famous archaeologist Prof. Lambda discovered an ancient Greek inscription which recorded that a foreign prince, whose name was given as Κανδείκλης, came with his followers into Greece and established his tribe there. The Professor anxious to identify the prince, after exhausting other sources of information, began to look through the letters C and K in the *Encyclopaedia Athenica*. His attention was attracted by an article on Canticles who it appeared was the son of Solomon. Clearly that was the required identification; no one could doubt that Κανδείκλης was the Jewish Prince Canticles. His theory attained great notoriety. At that time the Great Powers of Greece and Palestine were concluding an Entente and the Greek Prime Minister in an eloquent peroration made touching reference to the newly discovered historical ties of kinship between the two nations. Some time later Prof. Lambda happened to refer to the article again and discovered that he had made an unfortunate mistake; he had misread "Son of Solomon" for "Song of Solomon". The correction was published widely, and it might have been supposed that the Canticles theory would die a natural death. But no; Greeks and Palestinians continued to believe in their kinship, and the Greek Minister continued to make perorations. Prof. Lambda one day ventured to remonstrate with him. The Minister turned on him severely, "How do you know that Solomon had not a son called Canticles? You are not omniscient". The Professor, having reflected on the rather extensive character of Solomon's matrimonial establishment, found it difficult to reply.

The curious thing is that the determinist who takes this line is under the illusion that he is adopting a more modest attitude in regard to our scientific knowledge than the indeterminist. The indeterminist is accused of claiming omniscience. I do not bring quite the same countercharge against the determinist; but surely it is only the man who thinks himself *nearly* omniscient who would have the audacity to enumerate the possibilities which (it occurs to him) might exist unknown to him. I suspect that some of the other chapters in this book will be criticised for including hypotheses and deductions for which the evidence is considered to be insufficiently conclusive; that is inevitable if one is to give a picture of physical science in the process of development and discuss the current problems which occupy our thoughts. I tremble to think what the critics would say if I included a conjecture solely on the ground that, not being omniscient, I do not know that it is false.

I have already said that determinism is not disproved by physics. But it is the determinist who puts forward a positive proposal and the onus of proof is on him. He wishes to base on our ordinary experience of the sequence of cause and effect a wide generalisation called the Principle of Causality. Since physics to-day represents this experience as the result of statistical laws without any reference to the principle of causality, it is obvious that the generalisation has nothing to commend it so far as observational evidence is concerned. The indeterminists therefore regard it as they do any other entirely unsupported hypothesis. It is part of the tactics of the advocate of determinism to turn our unbelief in his conjecture into a positive conjecture of our own—a sort of Principle of Uncausality. The indeterminist is sometimes said to postulate "something like free-will" in the individual atoms. *Something like* is conveniently vague; the various mechanisms used in daily life have their obstinate moods and may be said to display something like free-will. But if it is suggested that we postulate psychological characters in the individual atoms of the kind which appear in our minds as human free-will, I deny this altogether. We do not discard one rash generalisation only to fall into another equally rash.

IV

When determinism was believed to prevail in the physical world, the question naturally arose, how far did it govern human activities? The question has often been confused by assuming that human activity belongs to a totally separate sphere—a mental sphere. But man has a body as well as a mind. The movements of his limbs, the sound waves which issue from his lips, the twinkle in his eye, are all phenomena of the physical world, and unless expressly excluded would be predetermined along with other physical phenomena. We can, if we like, distinguish two forms of determinism: (1) The scheme of causal law predetermines all human thoughts, emotions and volitions; (2) it predetermines human actions but not human motives and volitions. The second seems less drastic and probably commends itself to the liberal-minded, but the concession really amounts to very little. Under it a man can think what he likes, but he can only say that which the laws of physics preordain.

The essential point is that, if determinism is to have any

definable meaning, the domain of deterministic law must be a closed system; that is to say, all the data used in predicting must themselves be capable of being predicted. Whatever predetermines the future must itself be predetermined by the past. The movements of human bodies are part of the complete data of prediction of future states of the material universe; and if we include them for this purpose we must include them also as data which (it is asserted) can be predicted.

We must also note a semi-deterministic view, which asserts determinism for inorganic phenomena but supposes that it can be overridden by the interference of consciousness. Determinism in the material universe then applies only to phenomena in which it is assured that consciousness is not intervening directly or indirectly. It would be difficult to accept such a view nowadays. I suppose that most of those who expect determinism ultimately to reappear in physics do so from the feeling that there is some kind of logical necessity for it; but it can scarcely be a logical necessity if it is capable of being overridden. The hypothesis puts the scientific investigator in the position of being afraid to prove too much; he must show that effect is firmly linked to cause, but not so firmly that consciousness is unable to break the link. Finally we have to remember that physical law is arrived at from the analysis of conscious experience; it is the solution of the cryptogram contained in the story of consciousness. How then can we represent consciousness as being not only outside it but inimical to it?

The revolution of theory which has expelled determinism from present-day physics has therefore the important consequence that it is no longer necessary to suppose that human actions are completely predetermined. Although the door of human freedom is opened, it is not flung wide open; only a chink of daylight appears. But I think this is sufficient to justify a reorientation of our attitude to the problem. If our new-found freedom is like that of the mass of ·001 mgm., which is only allowed to stray $\frac{1}{5000}$ mm. in a thousand years, it is not much to boast of. The physical results do not spontaneously suggest any higher degree of freedom than this. But it seems to me that philosophical, psychological, and in fact commonsense arguments for greater freedom are so cogent that we are justified in trying to prise the door further open now that it is not actually barred. How can this be done without violence to physics?

If we could attribute the large-scale movements of our bodies to the "trigger action" of the unpredetermined behaviour of a few key atoms in our brain cells the problem would be simple; for individual atoms have wide indeterminacy of behaviour. It is obvious that there is a great deal of trigger action in our bodily mechanism, as when the pent up energy of a muscle is released by a minute physical change in a nerve; but it would be rash to suppose that the physical controlling cause is contained in the configuration of a few dozen atoms. I should conjecture that the smallest unit of structure in which the physical effects of volition have their origin contains many billions of atoms. If such a unit behaved like an inorganic system of similar mass the indeterminacy would be insufficient to allow appreciable freedom. My own tentative view is that this "conscious unit" does in fact differ from an inorganic system in having a much higher indeterminacy of behaviour—simply because of the unitary nature of that which in reality it represents, namely the Ego.

We have to remember that the physical world of atoms, electrons, quanta, etc., is the abstract symbolic representation of something. Generally we do not know anything of the background of the symbols—we do not know the inner nature of what is being symbolised. But at a point of contact of the physical world with consciousness, we have acquaintance with the conscious unity—the self or mind—whose physical aspect and symbol is the brain cell. Our method of physical analysis leads us to dissect this cell into atoms similar to the atoms in any non-conscious region of the world. But whereas in other regions each atom (so far as its behaviour is indeterminate) is governed independently by chance, in the conscious cell the behaviour symbolises a single volition of the spirit and not a conflict of billions of independent impulses. It seems to me that we must attribute some kind of unitary behaviour to the physical terminal of consciousness, otherwise the physical symbolism is not an appropriate representation of the mental unit which is being symbolised.

We conclude then that the activities of consciousness do not violate the laws of physics, since in the present indeterministic scheme there is freedom to operate within them. But at first sight they seem to involve something which we previously described (p. 64) as worse than a violation of the laws of physics, namely an exceedingly improbable coincidence. That had reference to coincidences ascribed to chance. Here we do not suppose that the conspiracy of the atoms in a brain cell to bring about a certain physical result instead of all fighting against one another is due to a chance coincidence. The unanimity is rather the condition that the atoms form a legitimate representation of that which is itself a unit in the mental reality behind the world of symbols.

The two aspects of human freedom on which I would lay most stress are *responsibility* and *self-understanding*. The nature of responsibility brings us to a well-known dilemma which I am no more able to solve than hundreds who have tried before me. How can we be responsible for our own good or evil nature? We feel that we can to some extent change our nature; we can reform or deteriorate. But is not the reforming or deteriorating impulse also in our nature? Or, if it is not in us, how can we be responsible for it? I will not add to the many discussions of this difficulty, for I have no solution to suggest. I will only say that I cannot accept as satisfactory the solution sometimes offered, that responsibility is a self-contradictory illusion. The solution does not seem to me to fit the data. Just as a theory of matter has to correspond to our perceptions of matter so a theory of the human spirit has to correspond to our inner perception of our spiritual nature. And to me it seems that responsibility is one of the fundamental facts of our nature. If I can be deluded over such a matter of immediate knowledge—the very nature of the being that I myself am—it is hard to see where any trustworthy beginning of knowledge is to be found.

I pass on to another aspect of the freedom allowed under physical indeterminacy, which seems to be quite distinct from the question of Free Will. Suppose that I have hit on a piece of mathematical research which promises interesting results. The assurance that I most desire is that the result which I write down at the end shall be the work of a mind which respects truth and logic, not the work of a hand which respects Maxwell's equations and the conservation of energy. In this case I am by no means anxious to stress the fact (if it is a fact) that the operations of my mind are unpredictable. Indeed I often prefer to use a multiplying machine whose results are less unpredictable than those of my own mental arithmetic. But the truth of the result $7 \times 11 = 77$ lies in its character as a possible mental operation and not in the fact that it is turned out automatically by a special combination of cog-wheels. I attach importance to the physical unpredictability of the motion of my pen, because it leaves it free to respond to the thought evolved in my brain which may or may not have been predetermined by the mental characteristics of my nature. If the mathematical argument in my mind is compelled to reach the conclusion which a deterministic system of physical law has preordained that my hands shall write down, then reasoning must be explained away as a process quite other than that which I feel it to be.

But my whole respect for reasoning is based on the hypothesis that it *is* what I feel it to be.

I do not think we can take liberties with that immediate self-knowledge of consciousness by which we are aware of ourselves as responsible, truth-seeking, reasoning, striving. The external world is not what it seems; we can transform our conception of it as we will provided that the system of signals passing from it to the mind is conserved. But as we draw nearer to the source of all knowledge the stream should run clearer. At least that is the hypothesis that the scientist is compelled to make, else where shall he start to look for truth? The Problem of Experience becomes unintelligible unless it is considered as the quest of a responsible, truth-seeking, reasoning spirit. These characteristics of the spirit therefore become the first datum of the problem.

The conceptions of physics are becoming difficult to understand. First relativity theory, then quantum theory and wave mechanics, have transformed the universe, making it seem fantastic to our minds. And perhaps the end is not yet. But there is another side to the transformation. Naïve realism, materialism, and mechanistic conceptions of phenomena were simple to understand; but I think that it was only by closing our eyes to the essential nature of conscious experience that they could be made to seem credible. These revolutions of scientific thought are clearing up the deeper contradictions between life and theoretical knowledge. The latest phase with its release from determinism is one of the greatest steps in the reconciliation. I would even say that in the present indeterministic theory of the physical universe we have reached something which a reasonable man might *almost* believe.

CHAPTER V

INDETERMINACY AND QUANTUM THEORY

That's Shell—that was!!
Well-known Advertisement.

I

WE have seen that all knowledge of physical objects is inferential. The external world of physics is a universe populated with *inferences*. Familiar objects which we handle are just as much inferential as a remote star inferred from an image on a photographic plate or an "undiscovered" planet inferred from irregularities in the motion of Uranus.

In the universe of inferences past, present and future appear indiscriminately, and it requires scientific analysis to sort them out. By a certain rule of inference, viz. the law of gravitation, we infer the present or past existence of an invisible companion to a star; by an application of the same rule we infer the existence on Aug. 11, 1999, of a configuration of the sun, earth and moon which corresponds to a total eclipse of the sun. In principle we have no reason to place greater confidence in the present inference than in the future inference; indeed it would generally be considered that we are less likely to have made a mistake about the eclipse. Both are of the same nature as the familiar inference that there is —or was!!—a motor car "over yonder", which depends on our experience that light generally travels in straight lines, a law which is by no means so regularly obeyed as the law of gravitation. The shadow of the moon on Cornwall in 1999 is already in the world of inference. It will not change its status when the year 1999 arrives and we observe the eclipse; we shall merely substitute one method of inferring the shadow for another. The shadow will always be an inference. By the shadow I here mean the entity or condition in the physical world, viz. a comparatively quiescent state of the aether, not the sensory perception of darkness in a number of human and animal minds.

Of particular importance for the problem of determinism are our inferences about the past. Strictly speaking our most direct inferences from sight, sound and touch all relate to a time slightly antecedent to the sensation. To obtain an inference as to the present state of things we have to combine them with our general inferential knowledge of the continuity of phenomena obtained from other experiences. But there are cases in which the time lag is more considerable, or for other reasons the argument of continuity does not apply. Suppose that we wish to determine the chemical constitution of a certain salt. We put it in a test tube and apply various reagents, and from the phenomena observed reach the conclusion that it *was* silver nitrate. It is no longer silver nitrate after our treatment of it. The property which we infer is not that of "being X" but of "having been X". We say in fact "That's X—that was!!" I will call this *retrospective inference*.

We noted at the outset (p. 76) that in considering determinism the alleged causes must be challenged to produce their birth-certificates, so that we may know whether they really were pre-existing. Retrospective inference is particularly dangerous in this connection because it involves antedating a certificate. The experiment above mentioned certifies the chemical constitution of a substance, but the date we write on the certificate is earlier than that at which we became assured of the composition.

To show how retrospective inference might be abused, suppose that there were no way of learning the chemical composition of a substance without destroying it. By hypothesis a chemist would never know until after his experiment the composition of the substance he had been handling, so that the result of every experiment must be unforeseen. Must he then admit that the science of chemistry is chaotic? A man of resource would override so trifling an obstacle. If he were discreet enough never to say beforehand what his experiment was going to demonstrate, he might give edifying lectures on the uniformity of Nature. He puts a lighted match in a cylinder of gas and the gas burns—"There you see that hydrogen is inflammable". Or the match goes out—"That proves that nitrogen does not support combustion". Or the match burns more brightly—"Oxygen feeds combustion". "How do you know it was oxygen?" "By retrospective inference from the observation that the match burns more brightly." And so the experimenter passes from cylinder to cylinder, and the match does now one thing and now another, thereby beautifully demonstrating the uniformity of Nature and the determinism of chemical law!

If by retrospective inference we infer causal characters at an earlier date and then say that those characters invariably produce at a future date the manifestations from which we made the inference, we are working in a vicious circle. The connection is not causation but definition, and we are not prophets but tautologists. We must not mix up the genuine achievements of scientific prediction with this kind of charlatanry, nor the observed uniformities of Nature with those so easily invented by our imaginary lecturer. If we are to avoid vicious circles we must refuse to recognise purely retrospective characteristics—those which are never found as existing but always as having existed. If they do not manifest themselves until the moment that they cease to exist they can never be used for prediction except by those who prophesy after the event.

Chemical constitution is not one of these retrospective

characters, although it is often inferred retrospectively. The fact that silver nitrate can be bought and sold shows that there is a property of being silver nitrate as well as of having been silver nitrate. If a property can be assigned retrospectively the method of sampling usually enables us to assign the same property simultaneously. We divide a given substance into two parts, analyse one part (destroying it if necessary) and show that its constitution *has been* X; then it is usually a fair inference that the constitution of the other part *is* X. If that method were universally applicable there would be no danger of introducing into physics characters which have only a retrospective existence. But the method of sampling is inapplicable when we consider those characteristics which are supposed to distinguish one atom from another; for the individual atom cannot be divided into two samples, one to analyse and one to preserve. So it is in the domain of atomic physics that the confusion caused by retrospective inference has arisen.

It is known that potassium consists of two kinds of atoms, one kind being radio-active and the other inert. Let us call the two kinds K_α and K_β. If we observe that a particular atom bursts in the radio-active manner we shall infer that it was a K_α atom. Can we say that the explosion was predetermined by the fact that it was a K_α and not a K_β atom? On the information stated there would be no justification at all; K_α is merely an antedated label which we attach to the atom when we see that it has burst. We can always do that however undetermined the event may be which occasions the label. When I see at Cambridge station an assemblage of parcels from different parts of the country all bearing the Cambridge label, I infer an efficient organisation. But that is on the supposition that the parcels were labelled when they were dispatched. It is no proof of organisation if someone has gone round sticking Cambridge labels on everything that happened to turn up at Cambridge. Actually, however, we have information which shows that the burst of the potassium atom is not undetermined. Potassium is found to consist of two isotopes of atomic weights 39 and 41; and it is believed that 41 is the radio-active kind, 39 being inert. It is possible to separate the two isotopes and to pick out atoms known to be K_{41}. Thus K_{41} is a contemporaneous character and can legitimately predetermine the subsequent radio-active outburst; it replaces the character K_α which was found retrospectively.

So much for the fact of outburst; now consider the time of outburst. Nothing is known as to the time when a particular K_{41} atom will burst except that it will probably be within the next billion years. If, however, we observe that it bursts at a time t we can ascribe to the atom the retrospective character K_t, meaning that it had (all along) the property that it was going to burst at time t. Now according to modern physics the character K_t is not manifested in any way—is not even represented in our mathematical description of the atom—until the time t when the burst occurs and the character K_t having finished its job disappears. In these circumstances K_t is not a predetermining cause. Our retrospective label adds nothing to the plain observational fact that the burst occurred without warning at the moment t; it is merely a device for ringing a change on the tenses.

The super-intelligence imagined by Laplace was able to foresee the whole future; but the proviso was that he must be acquainted with all the conditions prevailing *at a given instant*. How much does this proviso include? If it includes all retrospective characters that might be attributed, that is to say all that might be inferred by retrospective inference from what actually will happen in the future, he is using the future to predict the future. He can predict the exact time of break-up of the radio-active atom if he is told the character K_t of the atom; but that is just the same as being told the

INDETERMINACY AND QUANTUM THEORY

time of break-up. Clearly then we must exclude retrospective characters. And if Laplace's being cannot predict the future without them, we turn instead to the being whom secondary law has substituted, whose vision of the future is incomplete and nowhere reaches entire certainty but, so far as it goes, has the merit of being genuine foreknowledge.

We have seen that a retrospective character is a device for ringing the changes on the tenses. Such a device may be useful in systematising our knowledge. By replacing the undetermined events of the future by indeterminate characters ascribed to the present we telescope the whole course of events into one apparently instantaneous scheme. The indeterminism of the future is accordingly made to appear as an indeterminacy of the present. From the purely philosophical point of view this is a confusing way of expressing things; but that is of no particular concern to the scientist who is willing to adopt any device which helps him to get on with the job of formulating and applying the laws which decide the recurrencies of experience in an indeterministic world. In the next section we shall see how this device works.

II

In 1927 W. Heisenberg formulated an important principle which defines clearly the amount of indeterminism in the accepted system of physical law. It is called the Principle of Uncertainty or sometimes the Principle of Indeterminacy. This was not the origin of the change over of physics from determinism to indeterminism; but it called attention to the existing indeterminism in a way which could scarcely be overlooked even by those least attentive to the philosophy of science.

Laplace imagined an intelligence who "would include in one and the same formula" the movements of all the bodies in the universe. But to include them in a formula is not necessarily the same thing as to know them. An algebraic symbol may stand for a known or for an unknown quantity. So when we have a formula which professes to give exactly the future position of an object, the question arises whether it is given in terms of known or of unknown symbols. Heisenberg's principle tells us that *just half of the symbols represent knowable quantities and the other half represent unknowable quantities*. The unknowable quantities correspond to retrospective characters. By inventing such characters we make the future appear determinate; but they do not actually predetermine the future because they are themselves indeterminate until the future events have taken place.

This may seem a rather artificial way of describing the indeterminism of the future, but it shows that there is method even in indeterminacy. Looking first at the consequences of the indeterminacy we find that some phenomena are predictable with practical certainty whereas others are almost wholly spontaneous, but we do not discover any simple rule. But looking at the causes of the indeterminism (if an Irishism may be allowed) we find that what is lacking to secure a complete and certain prediction of the whole future is always *just half* of the total data that would be needed. The data are paired in such a way that for each datum or character inferable from manifestations up to a given instant there is a symmetrical datum—a retrospective character—which is not inferable until later; and without both data the exact prediction is impossible. In this sense the future is half linked to the past and half detached from it.

This is often expressed in the form that there is an *interference* between our experimental attempts to determine the two data. That has the disadvantage that it raises the question of our skill and ingenuity. I do not want to reopen the whole question of determinism *versus* indeterminism discussed in the last chapter; my purpose now is not to defend but to examine the implications of the indeterminism contained in the existing physical theory. According to that theory there is an incompatibility of the two data inherent in their own nature and not due to the operations of an intervening experimenter.

Let us consider an isolated system. It is part of the universe of inference, and all that can be embodied in it must be capable of being inferred from the influences which it broadcasts. Whenever we state the properties of a body in terms of physical quantities we are imparting knowledge as to the response of various external indicators to its presence and nothing more. A knowledge of the response of all kinds of objects would determine completely its relation to its environment, leaving only its un-get-at-able inner nature which is outside the scope of physics.* Thus if the system is really isolated, so that it has no interaction at all with its surroundings, it has no properties belonging to physics but only an inner nature which is beyond physics. So we must modify the conditions a little. Let it for a moment have some interaction with the world exterior to it; the interaction starts a train of influences which may reach the nerves and brain of an observer and become translated into sensory experience. From this one signal the observer can draw an inference about the system, i.e. he can fix the value of one of the symbols describing the system or fix an equation connecting several such symbols. To determine more symbols there must be further interactions resulting in sensory experience, one for each symbol fixed.

It might seem that in time we could fix all the symbols in this way, so that there would be no undetermined symbols in the description of the system and no unknown quantities in the equations which profess to foretell the future. But it must be remembered that the interaction which disturbs the external world by sending a signal through it also reacts on

* *The Nature of the Physical World*, p. 257.

the system. There is thus a double consequence; the interaction starts a signal informing us that the value of a certain symbol q in the system is q_1, and at the same time it alters to an unknown extent the value of another symbol p in the system. If we have learnt from former signals that the value of p is p_1, our knowledge ceases to apply, and we must start again to find the new value of p. Presently there is another interaction which tells us that p is now p_2; but the same interaction knocks out q, so that we no longer know its value. The observer is like the comedian with an armful of parcels; each time he picks up one he drops another.

It is of the utmost importance for prediction that the quantity which is upset by the interaction is not the quantity we are inferring but a paired quantity. If the signal taught us that at the moment of the interaction q was q_1 but that as the result of the interaction it has been changed to an unknown extent, we should never have anything but retrospective knowledge—like my imaginary chemistry lecturer who always destroyed his substances in the act of ascertaining their composition. The pairing allows us to have contemporaneous knowledge of half the symbols but never more than half. This, of course, is not an a priori rule of indeterminacy. Heisenberg's discovery was that it is the rule of the indeterminacy which applies to the physical universe.

Heisenberg's principle contains something more. We have been contemplating only two alternatives, viz. that the value of a symbol q is either known or unknown. But it may be partially known; that is to say, we may know it within certain limits of accuracy and with a certain degree of probability. If one of two paired symbols is known with certainty and accuracy the other must be altogether unknown; but if one is partially known the other may be partially known. For such partial knowledge Heisenberg's principle gives the rule that the uncertainty (or standard deviation) of the quantity q multiplied by the uncertainty of the paired quantity p is of the order of magnitude of Planck's constant h. The product of the two standard deviations is of the order of magnitude of one quantum.

The general definition of the paired symbols is rather a technical matter; they are called *coordinates* and *momenta*, each coordinate having a momentum paired with it. To show the results of the principle we will consider the position and velocity of an electron. We can fix the position of an electron to within about ·001 mm. and (simultaneously) the velocity to within about 1 km. per sec.; or we can fix the position to ·0001 mm. and the velocity to 10 km. per sec.; or the position to ·00001 mm. and the velocity to 100 km. per sec.; and so on. We can divide the uncertainty as we like, but it cannot be got rid of. The secret is that if by our experimental arrangements we persuade the electron to send us a very sharp signal of its position, its velocity (which it had previously signalled) is altogether upset by the reaction. But it is possible to compromise. If we allow the electron to send a less precise indication of its position, the reaction is less intense and the velocity does not get so bad a knock.

This combination of uncertainty is actually embodied in the present theoretical picture of an electron. Nowadays we represent an electron not by a corpuscle but by a packet of waves; and the notion of exact position coupled with exact velocity which applies to a corpuscle does not apply to a packet of waves. So if we describe something as having exact position and exact velocity, we cannot be describing an electron; just as (according to Bertrand Russell) if we describe a person who knows what he is talking about and whether what he is saying is true, we cannot be describing a pure mathematician. It is therefore not a question of lack of skill on our part, or a casual difficulty in the experimental handling of these minute objects, or a perverse delight of Nature in tantalising us. If ever the day arrives when by improved technique an experimenter measures the position and velocity of an electron with greater accuracy than Heisenberg's principle admits, the present quantum theory will join the limbo of forgotten theories.

We might spend a long time admiring the detailed working of these paired uncertainties which prevent us from knowing more than we ought to know. But I do not think you should look upon it as Nature's device to prevent us from seeing too far into the future. The future is not predetermined, and Nature has no need to protect herself from giving away plans which she has not yet made. But the mathematician has to protect his equations from making impossible predictions. It commonly happens that when we ask silly questions, mathematical theory does not directly refuse to answer but gives us an oracular answer like 0/0 out of which we cannot wring any meaning. Similarly when we ask where the electron will be to-morrow, the mathematical theory does not give the straightforward answer "It is impossible to say because it is not yet decided", because that is beyond the resources of an algebraic vocabulary. It gives us an ordinary formula in x's and y's, but it makes sure that we cannot possibly find out what the formula means—until to-morrow.

III

The Principle of Uncertainty has the same kind of position in physics as the Principle of Relativity. Both have arisen from the discovery of what appeared at first to be a tantalising limitation of our resources of observation. The theory of relativity originated in the discovery that we cannot observe the motion of ourselves or of anything else relative to the aether. That seemed at first to be a casual obstacle in our search for truth; but it is now realised that our failure was due to the fact that we were looking for something which did not exist. Since then we have been on the look out for other pitfalls of the same kind. We must make sure that the quantities or characters that we speak about are directly or indirectly definable in terms of experience—otherwise our words convey no meaning. It was suspected that something of this kind was at the root of the difficulties of the old quantum theory; but the precise point of failure of our definitions eluded detection. The following passage, written shortly after the great awakening brought about by the theory of relativity, will illustrate the thought of the time.*

I should be puzzled to say off-hand what is the series of operations and calculations involved in measuring a length of 10^{-15} cm.; nevertheless I shall refer to such a length when necessary as though it were a quantity of which the definition is obvious.... I may be laying myself open to the charge that I am doing the very thing I criticise in the older physics—using terms that have no definite observational meaning, and mingling with my physical quantities things which are not the results of any conceivable experimental operation. I would reply—By all means explore this criticism if you regard it as a promising field of inquiry. I here assume that you will probably find me a justification for my 10^{-15} cm.; but you may find that there is an insurmountable ambiguity in defining it. In the latter event you may be on the track of something which will give a new insight into the fundamental nature of the world. Indeed it has been suspected that the perplexities of quantum phenomena may arise from the tacit assumption that the notions of length and duration, acquired primarily from experiences in which the average effects of large numbers of quanta are involved, are applicable in the study of individual quanta. There may need to be much more excavation before we have brought to light all that is of value in this critical consideration of experimental knowledge. Meanwhile I want to set before you the treasure that has already been unearthed in this field.

The excavation has proceeded and has not revealed anything wrong with 10^{-15} cm., nor with very minute measurements of other quantities. The pitfall was just a stage more subtle than that which relativity theory had exposed. It is

* Eddington, *Mathematical Theory of Relativity*, p. 7 (1923).

the *combination of two exact measurements* which has proved to have no definable meaning in terms of experience, although either measurement alone would express something definite.

The remedy adopted by relativity theory was simple; it expelled the quantities such as "velocity through aether" which were found to have no meaning. Thus purged, the physical universe became identified with the knowable. But the same treatment could not be applied to two quantities which play "Box and Cox". We have had to give up the attempt to define an objective world which corresponds exactly to what is potentially knowable. We have instead a universe which is just half knowable, and we are free to choose which half we shall set about knowing. That at least is how it appears when described in terms of our ordinary epistemological outlook. Equivalently, by the substitution of two quantities partially known for one quantity known and the other unknown, we reach the outlook of wave mechanics. What is knowable, i.e. inferable from experience, is a distribution of probability; we infer, not a series of events in the objective universe but the degree of probability of all possible events in the objective universe. Thus between the universe inferable from experience and the objective universe there is interposed the rather baffling conception of probability, which we shall try to understand in the next chapter.

I have said that the indeterminism of the future applies to all phenomena, although for some it may be practically insignificant. Perhaps you will think this statement too sweeping. Referring again to the isotopes of potassium, it is not predetermined whether a million years hence a given atom of the radio-active isotope K_{41} will or will not have broken up. On the other hand K_{39} is non-radio-active and has not enough energy to explode. Then (it will be said) there is at least one predetermined fact about its future; we can predict without any indeterminism that a million years hence it will not have broken up. I am not going to object that you are pressing my statement unfairly. I meant just what I said—though I must ask you not to look on that as a precedent.* Strictly speaking there is no such thing as a K_{39} atom, but only an atom which has a high probability of being K_{39}. Such an atom should contain 39 protons within a comparatively small nucleus; but a proton in modern physics (like an electron) is never anywhere quite definitely though it may have a higher probability of being in one place than another. Thus we can never get beyond a high probability of 39 protons being collected together. It is impossible to trap modern physics into predicting anything with perfect determinism because it deals with probabilities from the outset.

IV

Heisenberg's principle has a very curious consequence when it is applied to an angular position and to a corresponding angular momentum. These are paired quantities, and the product of their uncertainties is a quantum, i.e. Planck's constant h. Consequently if we want to know the angular momentum of a system very accurately there must be a very wide uncertainty in our knowledge of the angular position or orientation of the system. But a difficulty arises. However careless we are, we cannot make a mistake of more than 360° in laying down an orientation, or—in the usual circular measure—the *greatest* possible uncertainty in angle is 2π. Therefore by Heisenberg's rule the *least* possible uncertainty in angular momentum is $h/2\pi$. This forms a kind of discrete unit of angular momentum. We cannot distinguish differences of spin finer than this unit. When we describe changes of spin of an atom such changes must amount to one or more units; for a smaller change has no meaning definable in terms of experience. We have thus to picture a kind of change

* See p. 279.

which can only occur by jumps of a whole unit. This is one way of realising the origin of the orbit "jumps" of an electron, which were such a mysterious feature of the older quantum theory. I daresay that viewed in this way they become even more mysterious; but at least they are now seen to be a special case of a very general principle which covers the whole indeterminacy of physics, and not merely a sporadic phenomenon inside an atom.

It also follows that in the small-scale systems we cannot separate geometry from dynamics. As soon as we introduce into our picture of the world anything possessing orientation, it automatically begins to spin one way or the other. To say that there is definitely *no* spin would be to claim an accuracy of knowledge which we have seen to be impossible. The most "restful" system we can contemplate is one equally likely to have any value of the spin up to half a unit in either direction. Knowing then the probability distribution, we can compute the average energy of spin of a large number of these restful atoms; in macroscopic physics the average is all that concerns us. The more complicated the system, the greater will be the number of directions or orientations defined in our picture of it; and since each of these has its own uncertainty of spin, there is on the average a considerable amount of angular momentum present which is of this irreducible kind. In short, in ascribing geometry to the system we are compelled by the Uncertainty Principle to ascribe to it energy of constitution. Observation may show us that more than this minimum energy is present. Such additional energy is called energy of *excitation*; it may be radiated or otherwise passed from system to system.

This kind of application of the Uncertainty Principle has been used by F. A. Lindemann* to explain a number of striking results and paradoxes of the quantum theory. I do not think we need trouble much about the rigour or

* *The Physical Significance of the Quantum Theory.*

INDETERMINACY AND QUANTUM THEORY 107

precision of the method. The Uncertainty Principle arises out of the wave constitution of electrons and protons; and those who put "safety first" will naturally proceed to these results by rigorous solution of the wave equations. In this book we are not prepared to follow the detailed progress of the mathematician, who is cautiously finding his way through the maze of passages in the edifice of wave mechanics; so we are grateful for a window through which we can catch a glimpse of one or two of the interesting rooms.

The guiding principle can perhaps be expressed as follows. It would be illogical to admit as a constituent of the external world a carbon atom whose properties were inconsistent with its being *known to be* a carbon atom—illogical because the name refers not to its inner nature (which is outside physics) but to its manifestations. Therefore when we speak of a carbon atom, we imply that it has undergone the reactions which are involved in signalling through its surroundings that it is such a system as a carbon atom is defined to be, namely a nucleus with six satellite electrons. Thus the mention of a carbon atom implies *inter alia* that it has been possible to count the number of electrons and make reasonably sure that the number is six not five. Let us imagine ourselves counting them: "One, two, three, four, five,—Now is that a sixth, or have I already counted it?" You cannot count unless the objects have some degree of fixity of position, as those who induce slumber by counting sheep in a green field are well aware. But the more closely you fix the positions the bigger the uncertainty of momentum. So when you consider six electrons in an atom you have to attribute (on the average) more angular momentum than is involved in connecting each individually to the atom. This is a consequence of introducing enough distinction of position for it to be externally manifest that six electrons are at work and not fewer. In this kind of way Lindemann arrives at the **Exclusion Principle** by which each electron in the atom must have a separate quantum orbit sufficiently differentiated from the orbits of the others.

My own interest in this method is bound up with its application to "finite but unbounded space". Here again the rigorous demonstration rests on the equations of wave mechanics; but the Uncertainty Principle is useful for a preliminary insight. We have seen that in the theory of relativity space-time has a natural curvature, so that three-dimensional space curves round and closes up analogously to the two-dimensional surface of a sphere. It is evident that in a finite space of this kind we cannot make so big a mistake about the position of anything, as we could in infinite open space. It is impossible to be more than 12,000 miles out in locating an inhabitant of the earth; and similarly there is an upper limit (some thousands of millions of light years) to the possible error in locating an electron or any other inhabitant of our more or less spherical universe. Just as before, this upper limit to the uncertainty of position implies a lower limit to the uncertainty of momentum; so that in the most favourable case, when we know nothing at all about the position of the electron except that it is *somewhere* (i.e. within the limits of the universe), it must have a small uncertainty of momentum. A large number of electrons and protons will accordingly possess a certain irreducible average momentum and energy.

This result of wave mechanics throws further light on the meeting point of relativity theory and quantum theory (p. 48). In relativity theory mass (or energy) and momentum are associated with curvature of space-time, and indeed are identified with the measures of certain components of the curvature; the law of conservation of energy and momentum and the gravitational effect which one mass exerts on another are deducible from this identification. On the other hand quantum theory has treated the energy and momentum of a particle empirically without revealing that they have any connection with curvature. We now see that energy and momentum will arise out of curvature of space according to the principles of quantum theory as well as according to the principles of relativity theory. The *modus operandi* is that curvature limits the extent of space available for the particle to roam over and so limits our ignorance of its position; hence by Heisenberg's principle there is introduced a minimum uncertainty and therefore a non-zero average value of the energy and momentum.

The further development of this theory of the origin of mass must be postponed to Chapter XI.

CHAPTER VI

PROBABILITY

It is remarkable that a science which began with the consideration of games of chance, should have become the most important object of human knowledge.   LAPLACE, *Théorie Analytique des Probabilités.*

I

ABOUT the beginning of the nineteenth century the mathematical theory of probability attained great prominence through the writings of Laplace, Gauss and other famous mathematicians. It has had many applications in physical science. At first it was almost wholly confined to the treatment of errors of observation—especially in astronomy, which seems to have enjoyed the doubtful distinction of being the subject which provides most scope for a theory of errors. With the rise of thermodynamics and the analysis of matter into great numbers of independent particles moving at random, probability has entered more intimately into the fundamental problems of physics. To-day the pre-eminent symbol in wave mechanics, the mysterious ψ which the quantum physicist pursues from equation to equation, is—in so far as we may define the indefinable—identified with probability. In the most modern theories of physics probability seems to have replaced aether as "the nominative of the verb 'to undulate'".

Since it is so often necessary to refer to probability in these lectures, I have thought it well to devote a chapter to this much debated subject. We ought at least to clarify our ideas sufficiently to use the conception of probability consistently and logically in its scientific application.

When a word in everyday use is adopted as an exact

scientific term it does not always retain its everyday meaning. For example, in mechanics *work* is a technical term having a meaning by no means coextensive with our ordinary notion of work. Scientifically no work is done unless something is moved. The acrobat who stands at the base of a tableau, with the other members of the troupe supported gracefully on his shoulders, does no work. Similarly it must not be expected that probability when used as an exact term in mathematics and physics will retain all the shades of meaning that it may have in ordinary conversation. As a technical scientific term it denotes something to which a definite numerical measure can be attributed; to secure this definiteness we must sacrifice some of the looser implications of probability.

Before proceeding to the scientific and mathematical definition let us examine the most common use of the word. We speak of the probability that a prisoner is guilty, or the probability that a certain course of action will be successful. The probability is rated as "high" or "low", but there is not usually any ground for assigning a numerical measure to it.

In this case probability refers to the strength of our expectation or belief. The probability of an event refers to the strength of our expectation that it will occur; the probability of a theory refers to the degree of confidence that a right-thinking person would have in it. I do not think there is any difference of substance between the two statements: (*a*) on the evidence it is highly probable that the prisoner is guilty, and (*b*) a right-thinking person would form from the evidence a strong belief that the prisoner is guilty. It must always be recognised that, both in the ordinary and in the scientific use of probability, the probability is dependent on or "is relative to" the information supplied; for additional information is likely to modify our expectation of an event or our confidence in a belief. In no circumstances is probability an absolute attribute of an event or a belief.

The question arises whether we can use the strength of

belief as a measure, or as the basis of a measure, of the probability. In my view this is impossible. At any rate the measurement of probability employed in mathematics and physics has an altogether different basis, as we shall see.

One difficulty in employing strength of belief as a measure of probability is that an expectation or belief has partly a subjective basis. We have agreed that it depends (and ought to depend) on the information or evidence supplied; but in addition the strength of the expectation depends on the personality of the man who weighs the evidence. We try to remove this subjective element by saying that the true probability corresponds to the judgment of a "right-thinking person"; but how shall we define this ideal referee? We do not mean a perfectly logical person, for there is no question of making a strictly logical deduction from the evidence; if that were possible the conclusion would be a matter of certainty not probability. We do not mean a person gifted with second-sight, for we want to know the probability relative to the information stated and not relative to occult information. We do not particularly mean a person of long experience in similar judgments, for he is likely to use his past experience to supplement surreptitiously the information on which the judgment of probability is ostensibly based. Apart from the obvious definition of a right-thinking person as "someone who thinks as I do" (which is probably the definition at the back of our minds) there seems to be no way of defining his qualities.

There are, of course, occasions when all sensible persons agree in rating the probability of one event as high and of another event as much lower; so that, if we do not attempt too precise a classification, the question of subjectivity of judgment does not arise. But there is no reason to think that these probabilities can be graded systematically in order of magnitude. It has been maintained by some writers that probability always has a numerical measure even when the word is used in this elementary way; and that the beliefs of a right-thinking person could ideally be arranged in a unique sequence in order of intensity. I rate this on a level with the view that to a person with a right sense of humour all jokes can be arranged in a unique sequence in order of funniness.

We conclude then that the most elementary use of the word probability refers to strength of expectation or belief which is not associated with any numerical measure. There can be no exact science of these non-numerical probabilities which reflect personal judgment. They form an important element in our outlook as do many other things which do not come within the scope of exact measurement. We act on such probabilities, and we are justified in so acting. Man is not just a logic factory. He is an adventurer, and the taking of risks is a condition of life. Expectations are sometimes fulfilled and sometimes disappointed. But Man goes on expecting.

II

We turn now to probabilities which admit of numerical measurement. Numerical estimates of probability are often made in ordinary conversation; e.g. "It is 5 to 1 that the prisoner is guilty". Here the intention is to give a general impression of the strength of one's belief, but no coherent explanation can be given as to why the measure number 5 was selected. Sometimes an expression of this form does not really refer to anything that could properly be called probability but concerns a proposed financial transaction. But other examples can be given in which the numerical probability has been calculated in a systematic way, and we are guided by these in formulating the scientific definition of probability. As an example of a probability whose measure is definite and commonly recognised we take the statement "The probability is $\frac{1}{6}$ that my next throw with the dice will be an ace".

Let us first find out precisely what this statement means. Like many common statements the meaning is not to be discovered by examining the grammatical structure of the sentence. The best way of realising the meaning is to consider what evidence we should accept as proving or supporting the statement. Ostensibly it is a statement about "my next throw"; it would therefore seem natural to test its truth by making my next throw. But it is well known that that would provide no evidence one way or the other. The ace may turn up, or it may not; in either case there is no reason to change our opinion as to whether the odds against it were correctly stated.

A recognised test would be to throw the dice 6000 times. If the number of aces thrown is reasonably near 1000, that is regarded as satisfactory confirmation that the probability is $\frac{1}{6}$. If it is, say, 1230, that is an almost conclusive disproof.

Thus, although the statement refers to my next throw, its meaning is not specially connected with my next throw.

Verbally the statement refers to a particular event; but its meaning refers to a class of events of which the particular event is one member. Thus numerical probability is a communal property, acquired through membership of a class. The statement—

The probability is p *that an event* a *has an outcome* e

has to be translated

The event a *is a member of a certain class of events* A, *and the proportion of events in the class* A *which have an outcome* e *is* p.

The proportion of events in a given class which have an outcome *e* is generally called the frequency of *e* in that class. Thus a numerical measure *frequency* belonging to a class is verbally transferred to an individual member of that class and renamed *probability*.

By this definition we introduce a probability which is not

based on strength of belief; it denotes simply the proportion of events with a given outcome in a defined class. We do not say that there is no connection between this kind of probability and strength of belief; for the frequency of success will, like any other relevant information, be taken into account in forming a belief in the rather indefinite way in which beliefs are formed. Certain beliefs may be mainly, or even wholly, based on a numerical probability. But there is no mathematical connection between the probability and the belief, for the passage from evidence to belief is not along mathematical lines.

We have examined two common uses of the word probability, the one a non-numerical probability associated with strength of belief or expectation, the other a numerical probability associated with frequency in a class. Both are well established in the language, and we can scarcely forgo either of them. Both may be introduced in a single sentence. If the dice have not been tested we are not sure that they are true, and therefore we are not sure that the frequency of an ace turning up is $\frac{1}{6}$. According to circumstances we may rate it as rather probable, highly probable, nearly certain, etc., that the dice are true. Then our statement will be "It is rather probable that the probability of my throwing an ace at the next throw is $\frac{1}{6}$". Here the first is a non-numerical probability referring to the strength of belief; the second is a numerical probability or frequency. The second is the probability that has been taken over into science where it is used as a technical term; but the scientist cannot monopolise the language, and he must at times also use the word with the other non-technical meaning.

Failure to distinguish the two usages has often caused obscurity in treating the subject. The common idea is that, since probability signifies uncertainty, a statement like the foregoing which contains two uncertainties ought to be reducible to simpler terms. But numerical probability is not an uncertainty; it is an ordinary physical datum—the frequency of a certain characteristic in a class. Our knowledge of it may be uncertain, but so too is our knowledge of many other physical data. The statement "It is rather probable that the probability is..." is no more objectionable than the statement "It is rather probable that the solar parallax is...".

Normally the class of events A consists of, or at least includes, events which have not yet occurred, and the frequency of the outcome e is deduced from theory and not from actual statistics. This theoretical information about the class is not furnished by the theory of probability. For example, certain operations such as shuffling are supposed to give certain results with equal frequency. Again, it is often assumed that certain events will in the future occur with the same frequency as they have been observed to do in the past. The study of probability is often distracted by a discussion as to whether we have any proof of these assumptions. But the function of probability theory is to utilise such information, not to supply it. When once it is realised that there is nothing illogical in a numerical probability being itself only probable, we can utilise any reasonable belief as to the frequency of events and so determine "a reasonably probable probability"; just as we may use a reasonable belief as to the cause of the recession of the spiral nebulae and so determine a reasonably probable cosmical constant.

You will see that I do not discuss why, after having ascertained that an event belongs to a class containing 9 successes to 1 failure, we generally form a fairly confident expectation that it will occur. I do not think this can be discussed apart from the formation of expectations based on other types of information. This is no doubt an important aspect of the subject of probability, but it is scarcely within our province. If we maintained, as some have done, that scientific (numerical) probability is the basis of all rational belief other than strict logical deduction,* thereby annexing the whole subject of inference to the mathematical theory of probability, it would be necessary to go into the matter further. But that is not the position here adopted.

III

We have seen that the probability assigned to an event is a property of a class of events. Usually the class is not directly mentioned in our statement; but there must be an implicit understanding, since otherwise the probability would be indeterminate. Thus I would say that the probability that Mussolini was born on a Friday is $\frac{1}{7}$; the understanding is that his birth is assigned to the class of all human births, and I believe (though I may be mistaken) that human births are equally frequent on all days of the week. You may have looked up the date and found it to be, say, Tuesday; if so, you will assign it to the more limited class of human births which have occurred on a Tuesday, and say that the probability is 0. We are both right. The probability relative to the information in my possession is $\frac{1}{7}$; relative to the greater information in your possession, it is 0.

This shows how probability comes to be relative to the information supplied. The information is used to define the class to which the event in question is assigned; additional information causes us to re-define the class. In this way more than one probability may belong to the same event. What is the probability that it will rain to-morrow (April 19)? This may refer to the frequency of rain on April 19 in all years; or to the frequency with which meteorological conditions similar to those now prevailing are followed by rain on the next day; or to a class satisfying both conditions.

* Logical deductions can be regarded as a special case corresponding to probability 1, i.e. certainty.

There are three or more numerical probabilities attached to the same event—a quite permissible situation.

The question arises which of these is the practical probability—the one by which we should be guided when we stand hesitating by the umbrella stand. If the probabilities were *certain* probabilities and not merely *probable* probabilities, there is no doubt that the third should be chosen—the one which embraces all the available information. This may be seen in the following way. Suppose that the frequency of rain on April 19 is quite definitely $\frac{1}{3}$. A man might bet 2 to 1 against its raining; and if he repeated the offer year by year he should come out even in the long run,* provided that he can always find someone to take his bet. But another man who took into account the information derived from weather forecasts could win money off him by accepting the bet only in those years when rain was predicted. Ideally then the probability on which we should act is the one which is relative to all the information obtainable; that is to say, the implied class A consists of events which are like a in every particular stated. But this only applies when the probabilities are known with reasonable certainty. Often they are somewhat uncertain generalisations based on limited past experience. Each additional piece of information cuts down the size of the class and thereby makes the generalisation more unsafe. Information which we have theoretical reason to believe is irrelevant, e.g. whether April 19 does or does not fall in Easter week on the occasion in question, should be excluded; it only does harm by cutting down the size of the class. The question to be settled is then, whether it is better to act on a very uncertain probability based on more information or a fairly certain probability based on less information. This is not the sort of question to be solved by mathematics.

* Subject to a growing fluctuation which the persistent gambler must be presumed to have decided to risk.

Naturally a mathematical theory can take no account of the uncertainty of the entities with which it deals, whether these entities be probabilities or other numerical quantities. By uncertainties I here mean those arising from dubious postulates, generalisations, etc.; measurable uncertainties, such as probable errors, can be (and should be) dealt with mathematically. The dilemma of having two differently computed probabilities to choose from is no different from that which arises in regard to many other physical quantities. By one method we determine the value of a physical constant with very great accuracy, except that there is a doubt whether the theory underlying the method is sound; by another method we obtain a much less accurate value, but we have more confidence in the theory on which it is based. To decide which result should have greater weight in determining our belief is the kind of job which we have earlier assigned to a hypothetical "right-thinking person". The mathematician declines to be a candidate for the post.

The objection to reducing the size of the class and thereby making generalisation more unsafe applies to the probabilities of everyday life and especially to those based on accumulated statistics, but it does not affect probabilities (frequencies) which are computed by pure theory. These are to be treated as definite probabilities not as merely probable probabilities. I do not mean that the theory is certainly true; but it is assumed to be true as the basis of discussion, and it is recognised that our results are contingent on the theory being right.* The theory will determine the frequency in a narrow class as definitely as in a wide class; there is therefore no disadvantage in cutting down the class, and we incorporate in the probability every scrap of information available.

* For example, one would not compute the probable position of a planet on the basis that Einstein's theory has a probability of $\frac{3}{4}$ and Newton's theory a probability of $\frac{1}{4}$. If the result is to have any scientific usefulness the computer must commit himself to one theory or the other.

Suppose, for example, we are considering the probability that an atom has a velocity within certain limits. We start with an initial class commonly supposed to be the class of all atoms from which, in the entire absence of information, we might suppose our particular atom to have been selected at random; the frequency of the given velocity in this class is called its a priori probability. I will not stop now to discuss this initial class, because it is as it stands a hopelessly illogical conception; and a critical study of how it is to be placed on a proper footing has very important consequences (p. 130). Next a variety of information is to be incorporated. The atom is in the earth's atmosphere; it is at a certain temperature; it has just undergone a collision with an α particle; it is an oxygen atom; and so on. Each piece of information as we introduce it cuts down the class by eliminating all those members which were inconsistent with it. Finally, after each new piece of information has had its whack at the diminishing class, we calculate the frequency of the given velocity in the class that remains; that then is the required probability—relative to all known information.

It is to be noticed that the information is used to define what is excluded, not what is included. Events incompatible with one of our items of information are excluded; events which are consistent with it are not necessarily included, because they may be contradicted by another item. This Exclusion Method is the only systematic way in which we can incorporate a number of separate observational results. I think it is very suggestive of the difference between the scientific and the familiar outlook. Ordinarily we expect our senses to tell us what there is in the external world; the scientist uses them rather to assure himself of what is *not* there. That is to say, he forms as wide a conception of the possibilities as he can, and tries to narrow them down by crucial experiments. His ideal is to state his conclusions about the external world in a sufficiently general form to include all possibilities that he is unable to give good reason for rejecting.

To illustrate this procedure by exclusion, I recall a question once set in a mock examination paper. It is true that it refers to the probabilities of everyday life instead of to the definite probabilities occurring in scientific theory. But in an examination paper the probabilities of everyday life become definite—for no candidate may doubt information that is vouched for by an examiner. The question was—

If A, B, C, D each speak the truth once in three times (independently), and A affirms that B denies that C declares that D is a liar, what is the probability that D was speaking the truth?

It was many years after I first heard of it that it occurred to me that the problem actually had an answer, and moreover was an instructive example of the Exclusion Method—the modification of the a priori probability first stated, by excluding those members of the class which are inconsistent with the additional information furnished.* The reader will be in a better position to appreciate the enormous advantage of the exclusion method if he has first been driven wild by attempting to solve the problem without it.

The difficulty in using the exclusion method is to obtain a start. The method provides for the addition of knowledge to knowledge, but not for the addition of knowledge to ignorance. Added information is used to narrow down the class of events contemplated; but, starting with complete absence of information, how do we obtain the initial class to be narrowed down by our first piece of information? In

---

* The combinations inconsistent with "*A* affirms, etc." are truth-lie-truth-truth and truth-lie-lie-lie, which occur, respectively, twice and eight times out of 81 occasions. Excluding these, D is left with $27-2$ truths to $54-8$ lies, so that the required probability is $25/71$. The solution, of course, does not pay heed to the psychology of the quarrel; e.g. we do not try to deduce anything from the fact that A was provoked to speak rather than to hold his tongue.

the foregoing problem of A, B, C and D, our first information specified the frequencies of a class so that no difficulty arose; but that was due to the benevolence of the examiner. Nature does not so kindly adjust her problems to our capacity. So the question has often worried us, What class of events corresponds to complete ignorance? The whole conception of such a class is a logical contradiction. The height of absurdity was reached in the much-discussed Principle of Indifference, which asserted that when there is no information all alternatives are equally probable. Heaven knows why! However, since there are an infinite number of ways of classifying alternatives, and the principle does not say which way is to be chosen, it leaves us none the wiser.

There are many instances in which it is plausible to assume that a number of alternatives are equally probable. It is not always easy to see that the plausibility rests on knowledge (or positive conjecture), never on ignorance. The statement that the probability is $\frac{1}{6}$ that my next throw will be an ace is only true in the sense originally intended if the dice are not loaded; but there is another sense in which the probability is still $\frac{1}{6}$ even if we suspect that the dice are loaded. The two interpretations are due to the class of events not being explicitly specified. In the first sense, the probability refers to frequency in the class of all throws with this particular cube; in the second sense, it refers to all throws with all dice. We presume that in the latter class, although loaded dice exist, they are loaded against all numbers equally. That is to say, we assume that the probability of a certain face of a cube bearing a given number, and the probability of its being the face nearest to the centre of gravity of the cube, are independent uncorrelated probabilities. This assumption involves some knowledge of the methods of making dice. We might easily argue against it. If the practice is to stamp the numbers in order, so that the number 1 is on the face which happened to be uppermost when the cube was picked up for stamping, the tendency will be for the loading to be in favour of the number 1. The actual practice may be altogether different, but my point is that the assumption of equal probability of throwing the six numbers is based on information, whether true or false, about the circumstances of manufacture. It is not true that the probability is $\frac{1}{6}$ if we have no information whatever; we must at least know that the usual process of manufacture is not that which I have described.

It will, I think, generally be found that when numerical probabilities seem to appear rather mysteriously out of ignorance, their actual basis is an assumption of non-correlation between different frequencies—an assumption which, whether justified or not by our knowledge of the circumstances, represents the belief on which we are relying when we assert the probability. The belief is positive. It is not adopted merely as the most non-committal solution of a problem presented by our ignorance. The strength of our belief that the actual circumstances are such as not to introduce correlation determines the strength of our belief in the consequent probability distribution.

IV

It is notorious that the theory of probability has often been applied fallaciously. The most common mistake is to neglect the interdependence of two or more probabilities and combine them by formulae which apply only to independent probabilities. As a rule the culprit is fully aware of the heinousness of such an offence; it is simply that he has not been alert enough to detect the interdependence. Many illustrations of this neglect of interdependence could be cited from scientific writings up to the present day; but I will choose an early example, of which an account is given in Bertrand's *Calcul des Probabilités*. Take warning then from the story of Condorcet and the Judges.

In the first days of its exuberance, the theory of probability was applied by some of the famous mathematicians Condorcet, Poisson, Laplace and others to proposals for minimising errors of justice. The great Laplace was responsible for extravagances scarcely less glaring than those I shall relate. The Marquis de Condorcet, who seems to have started the idea, was a prominent mathematician of the time. He considered the problem of securing that a man should run no more risk of being wrongfully condemned than he might be expected willingly to shoulder. Take, for instance, the proportion of those who were accidentally drowned at a certain crossing of the Rhône to the number who safely passed it. No one troubled about this risk; therefore they would cheerfully accept the same risk of being executed by mistake. By such considerations Condorcet decided that one miscarriage of justice in 144,768 trials was a suitable figure to aim at. He had assured himself that truly enlightened judges could be found who would deliver not more than one wrong judgment in five. Here the wonderful new theory came in. Take sixty-five such judges and require a majority of nine for a judgment against the prisoner, and the risk of a wrong sentence is reduced to the above figure.

It does not seem to have occurred to Condorcet that the truly enlightened judges might be to some extent guided by the evidence; and that when an innocent man is wrongly condemned it is usually because the evidence has seemed to point against him. His calculation had assumed that the right and wrong decisions of his sixty-five judges would be distributed independently of one another.

Condorcet was somewhat concerned lest, with the large increase of judicial posts, there might be insufficient judges of the same high standard. Still judges who made, say, one mistake in three could be used; it was only necessary further to increase their number. His only misgiving was that if, other classes being exhausted, it was necessary to include those who made more than one mistake in two, his method would break down. ("Not at all", adds Bertrand. "A sufficiently numerous assembly in which each member is wrong more often than not will certainly pronounce against the truth, and therefore give a sure means of knowing it—at least according to Condorcet's formulae.")

Eight years later the Revolution came. Well had it been for the Marquis de Condorcet could he have been assured of one truly enlightened judge. He died by his own hand to escape the tribunal.

V

Another source of fallacy is *inverse probability* or the probability of causes. In science the "causes" are usually alternative hypotheses or explanations. It is argued that if a certain observed result is 100,000 times more probable on hypothesis A than on hypothesis B, then hypothesis A is 100,000 times more probable than hypothesis B. In judging the credibility of the two hypotheses we should, of course, regard information of this kind as highly pertinent; but there is no justification for the inverse form of statement. Suppose that you take a penny from your pocket and, tossing it five times, note that it turns up heads each time. The chance of a sequence of five alike throws with a normal penny is $\frac{1}{16}$; with a double-headed penny the chance is unity. But you would not argue that it is 16 times more probable that your penny is double-headed than that it is normal.

We must not, however, forget that probability is always relative to the information supplied. In rejecting the argument that the penny is most probably double-headed, we use our secret information that double-headed pennies are rare. Setting aside that and all other information which is not openly stated, we should have had no particular reason to reject the probability of 16 to 1 as being the probability relative to the given information. On the other hand there is no reason to accept it. When the argument is examined in detail, it is found to assume that, prior to the tossing, it was equally likely that the penny we had got hold of was double-headed or normal. Even if this were true, it is just as illegitimate for the defender of inverse probability to use his secret information to this effect as it is for us to use our secret information to the contrary. The deduction of the probability of causes from the probability of their consequences is a game whose rules are such that no one can take part in it without cheating.

But how are we to get on in physics without inverse probability? All our knowledge of the external world is an inverse inference—an inference of cause from effect. We experience sensations and we attribute them to more or less probable causes existing in the external world. Let it first be said that as regards the general scheme of physical law inferred from our experience—the accepted key to the cryptogram—we do not attribute to it any numerical probability. The evidence appears to us to warrant a strong belief in it, and that is all we can say. But as regards observed individual features, we commonly state our conclusions as numerical probabilities. We measure the parallax of the star Capella, and infer a probability of $\frac{5}{6}$ that Capella is between 13 and 16 parsecs away from us. This is really a statement of inverse probability, for the actual calculation is that the set of measurements that we have made is one which was (before we made it) five times more likely to occur if Capella is within these limits of distance than if it is outside them.

I think that there is a more logical way of expressing our detailed knowledge of the universe. The science of astronomy will not collapse if it turns out that we have made a wrong inference about Capella. We can never be sure of particular inferences; therefore we should aim at a system of inference that will give conclusions of which in the long run not more than a stated proportion, say $1/q$, will be wrong. Hence, instead of making the definite inference that Capella is probably between 13 and 16 parsecs away (probability $\frac{5}{6}$), we make the probable inference (probability $\frac{5}{6}$) that Capella is definitely between 13 and 16 parsecs away. By adopting the latter form we use a direct instead of an inverse probability and the logical difficulties of the former form are avoided.

Accepted observational knowledge of the universe is then a function of q—a series of maps becoming more and more detailed as q decreases. Thus in the map in which $\frac{99}{100}$ of the features are correct, we place Capella between 12 and $18\frac{1}{2}$ parsecs away. In the map in which $\frac{5}{6}$ of the features are correct, we place it between 13 and 16 parsecs; we can afford to be more precise in our statements as we become more reckless of their truth. The series of maps starts (at $q=$ infinity) with a map which is entirely correct—but unfortunately entirely blank; it ends (at $q=1$) with a map full of the minutest detail of which only an infinitesimal proportion is correct. What a philosopher is to make of these maps I will not venture to say; but the scientist affirms that some of the intermediate maps (say between $q=5$ and $q=20$) can be of considerable assistance to a sojourner in the universe who has to find his way about.

To see how this outlook avoids any question of inverse probability, we may refer again to the double-headed penny. You have tossed up the penny five times and it has fallen heads (or alternatively tails) every time. It is suggested that you should infer that the chances are 15 to 1 that it is a double-headed (or double-tailed) penny. That is clearly not the right inference. But suppose that you do not claim to be making the "right" inference, but to be applying a system of inference of facts (not of chances which would be meaningless in the circumstances) which will lead you wrong not more than once in 16 times. You may then boldly infer that the penny which has fallen heads or tails five times in succession is abnormal. We have secret knowledge that you will be wrong this time. But you will only be wrong as regards one penny out of every 16 that you try; for 15 out of 16 will assure you of their normality by exhibiting both faces. Thus your system of inference fulfils what you claim for it.

VI

Probability has intertwined itself round the roots of physical science. In thermodynamics, in quantum theory, and whenever gross matter is treated as an aggregation of a vast number of particles, the laws of chance are involved. Probability leavens the secondary scheme of physical law—the laws which are obeyed because it is "too improbable" that they should be broken. This application demands our special consideration.

We have had examples of two ways of utilising an observation. We can consider what may be inferred from that observation alone and calculate the probability attached to our inference; or we can consider how the new information contained in the observation modifies the probabilities which corresponded to the previous state of our knowledge. The second is the exclusion method discussed in Section III; it lends itself to more systematic treatment and is used throughout thermodynamics and quantum theory. In the modern form of quantum theory, known as wave mechanics, the exclusion method has been developed into a fine art. Each observation is treated as excluding a number of alternatives which had not been inconsistent with earlier knowledge and were accordingly represented as existing in the probability distribution or "fog" whose history is being traced.

Broadly speaking wave mechanics pictures a universe whose substance is probability, whereas classical mechanics pictures a universe whose substance is mass, energy, momentum, electric and magnetic force, etc. In wave mechanics we examine the way the probability moves about and redistributes itself; in ordinary mechanics we find the way mass, momentum and electromagnetic field move or are propagated. In the former the waves, which give the subject its name, are waves of probability; in the latter we treat sound waves, electromagnetic waves and gravitational waves. For brevity these may be contrasted as a universe of probability and a universe of entities. They are, however, both aspects of the same universe whose description involves both probabilities and entities. The difference in point of view is that in the first we attach entities (electrons, protons, photons) to the probabilities which we study; in the second we attach probabilities to the entities which we study—only from the nature of the entities treated in classical physics the attached probabilities are all practical certainties (p. 78). In macroscopic physics the variety lies in the entities—greater or lesser masses, greater or lesser field strength—the probabilities being all similar units; in microscopic physics the position is inverted and the variety lies in the probabilities, the entities generally being all similar units, e.g. electrons. It is therefore found to be more businesslike and practical to contemplate a distribution of probabilities; and the entities attached to them tend rather to drop out of sight in our calculations and deductions.

We have seen that in order to use the exclusion method it is necessary to start with an initial class; and wave mechanics accordingly starts with an initial or "a priori probability distribution" of the positions and velocities of the electrons or other entities. A priori probability is essentially an *unobservable*, for when we introduce observational knowledge we obtain a modified probability relative to that knowledge. We therefore seem led into the old fallacy of the principle of indifference in supposing that there can exist a probability relative to complete ignorance. This is a most unsatisfactory feature of wave mechanics when considered by itself; but the difficulty disappears when wave mechanics is combined with relativity theory. The a priori probability distribution is then regarded in the same way as other unobservables are regarded in relativity theory, e.g. a frame of space and time. For the purpose of representation we adopt an arbitrarily chosen frame of space and time; but our choice makes no difference in the end when we translate our results directly into terms of what can be observed. Similarly we can adopt an arbitrarily chosen distribution of initial probability; our choice makes no difference in the end when we translate our results directly into terms of what can be observed.

Thus the initial probability referred to in quantum theory and in the kinetic theory of gases is not an a priori probability in any metaphysical sense. It is part of an arbitrarily chosen reference system; and it is no more necessary to decide whether one distribution of initial probability rather than another is the true inference from complete ignorance than it is to decide whether the yard or the metre is the true standard of length. We adopt any convenient a priori probability distribution as we adopt any convenient frame of space and time; but it follows from the unobservable character of these comparison systems that the laws of physics must be invariant for all transformations of them. The recognition of this invariance is another of the important steps in the unification of relativity theory and quantum theory.*

When you make a change of your system of reference, whether it be a change of the frame of space and time or of the initial probability distribution to which all observational information is applied, you must carry through the change to the bitter end. If you change your space-time frame in mechanics you must change it also in optics, otherwise you will reach erroneous conclusions in regard to an experiment, such as the Michelson-Morley experiment, in which both optics and mechanics are involved. Just as in former days the Michelson-Morley experiment was misunderstood through

* Previous steps are discussed on pp. 48, 108.

segregating the optical and the mechanical (or metrical) factors in the experiment, so at the present time our experiments on atoms and electrons are very generally misunderstood through segregating the microscopic (quantum theory) and macroscopic (relativity theory) factors in the experiment. In particular if, in considering an experiment on an electron, you change the adopted a priori probability distribution of position and velocity, you must consider the consequences of that change not only on the formulae describing the behaviour of the electron itself but on all the particles that make up the apparatus used in the experiment. For the result of the experiment is affected just as much by a change of behaviour of the apparatus as by a change of behaviour of the electron.

Since physics has been divided into two branches, quantum theory and relativity theory, the electron being studied by the former and the gross matter of the apparatus by the latter, the experiment has been under the charge of two partners neither of whom knows what the other is doing. Relativity, dealing with matter and field on the gross scale, treats of the averages associated with vast numbers of particles. Leaving aside electrical characteristics, it treats especially of the average energy of the particles and the associated quantities, average momentum and average stress-system; these are grouped together to form what is called an average energy-tensor. So the a priori probability distribution in quantum theory is represented in relativity theory by its average energy-tensor. But when it enters into relativity theory it receives a new name; it is called the fundamental or metrical tensor ($g^{\mu\nu}$). This is the characteristic of space (or aether) which determines what will be the measure of the distance of two specified points or of the interval of time between two specified events.

Now let us return to the quantum theory side of the partnership. The quantum physicist is studying, let us say, a system of two or three electrons whose positions he has temporarily denoted by certain symbols called coordinates. But he cannot tell what these symbols mean in an observational sense—he cannot tell what are the distances between the particles—without appealing to his partner to furnish him with the metrical tensor $g^{\mu\nu}$ which constitutes the code for translating the symbols into distances. Just as in an actual experiment he would have to borrow gross apparatus belonging to macroscopic physics to measure the distances, so in the theory he has to borrow a macroscopic tensor to calculate them. So he borrows the tensor $g^{\mu\nu}$ from the relativity physicist. What he generally fails to recognise is that this is simply *the averaged characteristics of his own a priori probability distribution being handed back to him.*

Unconscious of this identity the quantum physicist applies the metric $g^{\mu\nu}$ to the positions and directions of motion of his particles and hence introduces it into his description of their probability distribution. In particular, it is with reference to this metric that he describes the initial a priori probability distribution—the framework into which observational knowledge is incorporated by the exclusion method. He discovers a remarkable result! He finds that the initial probability distribution must be uniform and isotropic throughout space-time. This is not very surprising when we recall that the initial probability distribution furnishes the metric $g^{\mu\nu}$ which is then employed to measure the initial distribution.

I imagine him turning on me and saying—"You were wrong when you said that I was free to choose the initial a priori probability distribution arbitrarily. Nature has chosen a uniform and isotropic distribution, and forces her choice on me. Any other choice would not lead to the equations which are verified by experiment". I might reply by reminding him, that by the way in which it is used in connection with observational knowledge the initial distribution is necessarily outside observation, so that there must be a fallacy in his conclusion. But the important thing is to see the source of the fallacy. By using $g^{\mu\nu}$ for the metric in his description, he is using the initial probability distribution to describe the initial probability distribution. However arbitrary it may be by extraneous standards, *compared with itself* it is necessarily exhibited as uniform and isotropic.

Those apparent laws of Nature which express uniformity and isotropy arise because we measure the world with apparatus which is itself part of the world. The measuring apparatus and that which is measured are constituted ultimately of the same type of elementary particles; so that any asymmetry of behaviour must appear on both sides of the comparison and be eliminated in all our measurements. I have explained elsewhere* how Einstein's law of gravitation, which states that the curvature of the world in empty space or aether is uniform and isotropic, arises in this way. The uniformity and symmetry of the a priori probability distribution is of similar character, and is in fact a closely related aspect of the same investigation.

Coming back to the general theory of probability which is the subject of this chapter, we have been concerned to show that probability is always relative to knowledge (actual or presumed) and that there is no a priori probability of things in a metaphysical sense, i.e. a probability relative to complete ignorance. We have examined what at first appeared to be two cases of exception. In the example of the loaded dice, we have pointed out that what is assumed is not ignorance but knowledge that the circumstances of manufacture are such that two probabilities concerned in the problem are uncorrelated. The other example is the initial probability assumed in wave mechanics and other statistical branches of physics, which is commonly called a priori probability. It is, for example, assumed that the initial probability of finding a particle in a given region is simply proportional to the

* *The Nature of the Physical World*, pp. 138–145.

volume of the region; in other words all equal volumes have an equal amount of initial probability of containing the particle. This is often looked upon as an example of the Principle of Indifference—that initially (i.e. when no information is supplied) all alternatives are equally probable. But it has nothing to do with that principle. The proportionality of volume to initial probability is a physical law of precisely the same type as the proportionality of energy to mass. Such laws arise because there are two ways in which the same natural entity can affect our experience. Except that they are measured in different units mass is simply an *alias* of energy. Similarly the volume of a region is an *alias* of the initial probability of its containing the particle, the one name being used in macroscopic theory and the other in microscopic theory.

CHAPTER VII

THE CONSTITUTION OF THE STARS

> Study is like the heaven's glorious sun,
> That will not be deep-searched with saucy looks.
>
> SHAKESPEARE, *Love's Labour's Lost.*

I

THE history of exploration of the interior of a star begins in the year 1869 when J. Homer Lane wrote a famous paper entitled "On the Theoretical Temperature of the Sun, under the Hypothesis of a Gaseous Mass maintaining its Volume by its Internal Heat, and depending on the Laws of Gases as known to Terrestrial Experiment". He might perhaps have chosen a more snappy title. But the fullness has the advantage of bringing before us a number of important ideas. The various phrases each deserve close attention, and we shall use them as the firstly, secondly, thirdly, of our sermon. We shall consider other stars besides the sun, and other conditions of the interior besides the temperature; but everything centres on the problem of temperature. What is the degree of heat deep down inside these great celestial furnaces?

I would emphasise the phrase "depending on the laws of gases as known to terrestrial experiment". There is no speculative intention in these studies of the interior of a star. We simply want to find out how far the phenomena which we observe in the sky agree with and are a consequence of the laws that have been assigned to matter as the result of laboratory experiment. We encounter matter under conditions very different from those of the laboratory; and it may have something fresh to tell us—something quite unforeseen. But anything essentially new has to be sorted out from that which is a direct consequence of what we already know, or think we know. If the stars have any revolutionary ideas to suggest they will show it by a discordance from the results which we calculate for them on the basis of the accepted laws of physics.

Before diving into the interior, I must refer to our general knowledge of the stars as seen from outside. The number of stars within range of our most powerful telescopes is of the order of a thousand million, or say one apiece for every inhabitant of the earth. These, and many more stars too faint to be detected, form a great system which we call the Galaxy. This system is not the whole universe; but what lies beyond it will occupy us in Chapter x. The stars show a very wide diversity. Some are extremely dense and compact, others extremely tenuous. Some give out a million times as much light and heat as others. Some have a surface-temperature as high as 20,000° or perhaps 30,000°, others not more than 3000°. Some stars are believed to be pulsating, swelling and deflating with a period of a few hours or days. A considerable proportion occur in pairs—two stars revolving round each other. Some flock in clusters; the members of such clusters though widely separated from each other have at least the connection of a common origin. One star, we know, has a system of planets, and from one of those planets we view it; whether any other stars have such a system can only be guessed. According to Jeans there is theoretical reason to suppose that the evolution of a planetary system is a rare accident.

The most uniform characteristic of the stars is their mass, that is to say the amount of matter which constitutes them. A range from $\frac{1}{5}$ to 10 times the mass of the sun would cover all but the most exceptional objects. The general run of the masses is within a much narrower range. The most massive stars tend to force themselves on our notice because they are the most luminous; if we eliminate this selective effect, the

THE CONSTITUTION OF THE STARS 137

diversity of mass among a hundred stars picked at random would probably be not much greater than among a hundred men, women and children picked at random from a crowd.

When the spectroscope was applied to the detection of the various elements in the heavenly bodies, the first impression was that the stars varied greatly in chemical constitution. Elements prominent in the spectrum of one star are absent in another. Some stars, such as Sirius, show a spectrum which is almost wholly hydrogen; in the sun iron is very prominent; among the cooler stars, in which chemical compounds are not wholly dissociated by temperature, some indicate carbon compounds and others rather oddly specialise on titanium oxide. But these are not real differences of chemical composition. A particular spectrum can only appear if the physical conditions are such as are required to stimulate it. The variety of stellar spectra is therefore due primarily to the variety of physical conditions—differences of temperature and pressure in the layers which the spectroscope explores. We cannot be sure that the stars all have the same chemical composition; but if there are differences, it is by no means a straightforward problem to ascertain them.

In any case the composition of the layers bubbling on the outside of the stellar furnaces cannot be taken as a safe guide to the composition within. So we start our investigation of the stellar interior in practically complete ignorance of its chemical constitution. Leaving aside possible differences of chemical constitution, the stars may be expected to form a twofold sequence. We may specify a star by its mass and radius—the total amount of matter, and the space into which it is packed. We anticipate that these two data will fix all the other characteristics of the star—how much light and heat will pour out of it, what temperature its surface will take up, what will be the period of its pulsation if pulsation is possible, and so on. These presumably are necessary properties of a given amount of material forming a globe of given size, and it ought to be possible to calculate them by a study of the physical conditions. At any rate that is the line on which we may start to work, and if there are additional complications they will appear in due course.

It would be as difficult to select a "typical star" as to select a typical animal to represent the animal kingdom. But the sun is about as typical as any. It is not at all extreme in any of its characteristics; and around us there are numerous stars which are practically replicas of the sun. The sun is 330,000 times greater than the earth in mass and 1,300,000 times greater in volume. Its diameter is 865,000 miles, and its mass is 2000 quadrillion (2.10^{27}) tons. Its mean density is rather greater than that of water.

II

Viewing the sun from outside, we look down through the semi-transparent outermost layers. The level which is roughly the limit to which we can see down is called the photosphere. It is ascertained by fairly direct observational methods that the temperature at that level is nearly 6000° Centigrade. Continuing inwards below the photosphere the temperature must become higher and higher until it reaches its maximum at the centre of the sun; but we can only follow this increase by theory. It is found that by far the greater part of the interior mass is at a temperature above a million degrees. According to a favourite mathematical model the sun's central temperature is 21,000,000° and the mean temperature of the whole mass is 12,000,000°. These figures probably err in being, if anything, too high.

The clue which we follow in finding the internal temperature is contained in the title of Lane's paper—"a gaseous mass *maintaining its volume*". The mass of 2.10^{27} tons which constitutes the sun must exercise enormous internal pressure. If it were devoid of heat the matter would be crushed by this pressure into that strange condition which we find in the Companion of Sirius and other white dwarf stars, where the density is thousands of times greater than that of any material known on earth. Heat is required to distend the matter so as to occupy the actual volume of the sun. By heat we mean the energy of the random motions of the molecules. If the planets were deprived of motion they would all fall into the sun; so we may say that the solar system is kept distended by the motions of the planets. In the same way the sun is kept distended by the heat motions of the particles of which it is composed. By bringing together the various physical laws which bear on the subject, we have been able to make a tolerably close calculation of the amount of heat required to give the observed distension; and also to determine, but more roughly, how the heat must distribute itself through the sun in order to preserve a steady state.

The very high temperature has one effect which was not at first realised. There are two forms of heat—material heat, which is the energy of the particles, and radiant heat, which is the energy of aether waves. At terrestrial temperatures, for example in a white-hot mass of metal, the radiant heat is quite insignificant compared with the material heat. If we go near the white-hot mass we feel a great deal of radiant heat coming from it, but this is produced at the moment of emission by converting material heat in the iron; it is manufactured as required, and practically no reserve stock is kept. When the temperature is increased, the material heat increases roughly in proportion to the absolute temperature, but the radiant heat contained in the body goes up as the fourth power of the temperature, so that it gradually overtakes the material heat. Even at the temperature of the sun there is not so much radiant heat as material heat; and except possibly in a few of the most massive stars, the advantage is always with the material heat. But there is no longer any great disparity.

I have little doubt that it was this approximate balancing of the two forms of heat at some early stage of the evolution of a star that determined the standard mass to which the stars more or less closely conform. Something must have decided that the matter constituting our Galaxy has not all condensed into one mass but has divided into thousands of millions of stars, the majority of which are surprisingly uniform in the amount of material they contain. The mass of the pattern star cannot have been arbitrary. It seems significant that the mass is such that (especially in the earlier stages of condensation) the radiant heat is nearly on a parity with the material heat. For 50-fold greater or 50-fold smaller mass we should not have anything approaching a balance.

In Lane's discussion, and for a long time afterwards, the existence of this large quantity of radiant heat was not recognised. When the heat of the star was thought of as wholly material, it was necessary to postulate some means for bringing it up from the interior to the surface where heat is being radiated away. It was supposed that the matter at the surface cooled and sank down, and hot matter from below came up to take its place; so that throughout the sun there were convection currents bodily transferring the heat to the points where it had been lost and incidentally keeping the material well-stirred. But now the boot is on the other leg. Ever since it was recognised that the stars contain a vast quantity of radiant energy, the problem has been, not to devise a way of bringing heat up to the surface, but to understand how this highly mobile form of energy is held back from the surface—how it is encaged by the matter so that it does not leak out faster than we observe it to do.

At a temperature of some millions of degrees radiant energy consists of X rays. So in the stellar interior we have X rays in great abundance travelling in all directions. If the atoms and electrons in the sun were suddenly abolished the X rays now confined in the interior would scatter through space with the speed of light; 300,000 years' supply of radiation would be squandered in an instant. The atoms dam back this flood, catching and turning away the aether waves as they try to escape, absorbing and re-emitting them in a new direction, so that they go aimlessly round and round the maze. Thus only a slight leakage of radiation dribbles out to illuminate and warm the earth and the other planets.

This leads us to the principal aim of investigation of the stellar interior. Having found the internal distribution of temperature in the star and knowing therefore the quantity and quality of the radiant energy imprisoned there (for this is determined solely by the temperature), knowing also the density of the matter and therefore the number of atoms engaged in holding back the radiant energy, we ought to be able to calculate how much escapes. It is like calculating the flow of water through a pipe, when we know the head of water causing the flow and the resistance obstructing the flow. Here the increasing concentration of the X rays as the temperature increases inwards supplies the pressure gradient, and the opacity of the matter obstructing the transmission of X rays supplies the resistance. Our aim then is to calculate from the known laws of absorption of X rays how much radiation will get through, and so ascertain theoretically the amount of the energy flow which is slowly leaking outwards through the star. If our calculation is right, it will agree with the amount which emerges at the surface and constitutes the light and heat of the star. The calculation therefore gives us the brightness of the star, or more strictly the "heat-brightness" which measures the total radiant energy emitted irrespective of its luminous efficiency. The heat coming to us from many of the stars has been measured directly; but in any case the heat-magnitude of a star can easily be computed from its light-magnitude by applying a well-known correction depending on the colour or spectral type.

Effectively therefore by the method here outlined we

ought to be able to compute from the mass and radius of a star its theoretical luminosity. We can then compare this calculated luminosity with the luminosity observed. Before describing the results of this comparison there are a number of points to be considered.

As the escaping radiation travels from the hot interior to the comparatively cool surface layers, it is gradually transformed from X rays to longer wave-lengths or lower frequencies; so that the radiation which finally emerges consists of visual light together with some ultra-violet and infra-red rays as shown in the star's spectrum. The stepping-down of the frequency is automatic. Each unit of radiant energy—each photon—is being absorbed and re-emitted every few inches on its journey outwards; so that there is ample opportunity for adjusting the composition of the radiation to that proper to the temperature of the region which is being traversed. If we were to follow the last stages of the journey we should be concerned with the absorption of light instead of the absorption of X rays. But fortunately there is no need to trouble about this. Having, as it were, conducted the outflowing stream through $\frac{99}{100}$ of the material of the star, we can leave it to find its own way out. I say *fortunately*, because it is much easier to treat temperatures above a million degrees. It is the high temperature of the stars which makes our problem soluble. We could no doubt treat cooler matter in a similar way if we were given the necessary data. But we are not in a laboratory where we can find out any data we require; we are in the interior of a star provided with next to no data.

In this part of our discussion we are not concerned with the ultimate source of a star's heat. We take the star in the condition in which it now is and calculate that radiation is flowing through and out of it at a certain rate. Clearly there will be a gradual change in the condition of the star unless the heat inside it is being replenished from some source in the interior; and we infer that such replenishment occurs, because without it the star would change too rapidly for any admissible time-scale of evolution. But "rapid" here means perceptible in a thousand or a million years; it is not the kind of unsteadiness which would upset our calculation. So far as the calculation of the luminosity is concerned we do not care whether the star's store of heat is being replenished or not.

There is just one point at which our problem is not entirely detached from the problem of the source of maintenance of a star's heat. To obtain an exact result we should have to know how the source is distributed through the star—whether it is concentrated in the hottest central regions or is evenly distributed through the mass. We meet this difficulty by considering both extremes of distribution in turn, and calculating the luminosity on both hypotheses; the truth must lie between them. In this way it is found that the extreme uncertainty arising from our ignorance of the distribution of the source is for a typical star no more than $\pm 0^{m}\cdot 5$. We shall be well-satisfied if our calculation reaches this order of accuracy.

There can be little doubt that the heat of the sun and of other stars is being maintained by the liberation of some form of subatomic energy in the interior. Until the last two or three years the laboratory physicist had no information as to the conditions of release of subatomic energy; so on this side of the subject the astronomer could get no help from physics. Thus in developing the theory of the constitution of the stars depending on the laws of physics "as known to terrestrial experiment" progress was contingent on our being able to separate off an independent field of research which did not involve the unknown laws of subatomic energy. The problems treated in this chapter are segregated from the problem of subatomic energy in this way—except to the insignificant extent referred to in the last paragraph. Circumstances are now changing, and a great number of processes which release subatomic energy are being studied in the laboratory. It may be expected that before long important developments in the application to the source of stellar energy will follow. Some account of this side of the problem of stellar equilibrium will be given in Chapter VIII.

III

We have seen that an atom consists of a heavy nucleus together with a number of loosely attached satellite electrons belonging to it as the planets belong to the sun. By various kinds of maltreatment the physicist is able to detach one or more of the satellite electrons; this process of chipping off electrons is called ionisation, and the mutilated atom is called an ion. In laboratory conditions there is not much difficulty in producing a few ions among a great number of normal atoms; but when a certain proportion have been ionised, so that there are many homeless electrons wandering about, the ions are continually capturing these vagrants, and we soon reach a stage at which atoms are being made whole again as fast as we can ionise them.

At a temperature of 10 million degrees the forces of disruption are enormously intensified. Ionisation instead of being an occasional disease is epidemic. The intensification is in quantity rather than in quality. In a contest between the sun and the Cavendish Laboratory as to which could do the most violence to a single atom, I would back the Cavendish Laboratory. For the purpose of electrical experiments there is abundance of energy on the sun but very poor insulation. The efficiency of the sun is in mass-production. That helps us in our problem; for if mass-production is the only new feature, the multiplication table suffices to cope with it.

For definiteness consider an iron atom in the deep interior of the sun. Normally it should have 26 satellite electrons; 22 of these have broken loose and are wandering freely

THE CONSTITUTION OF THE STARS

through the material. Lighter elements such as carbon are stripped bare to the nucleus. Wandering electrons are always trying to settle in the vacant orbits and may succeed for an instant; but immediately an X ray comes along and explodes the electron away again. Perhaps I should add that the electron "wanders" at an average speed of 10,000 miles a second.

You can picture the commotion at 10 million degrees in the interior of the sun. Crowded together within a cubic centimetre there are more than a quadrillion (10^{24}) atoms, about twice as many electrons, and 20,600 trillion X rays.* We can speak of the *number* of X rays for the quantum theory gives them a kind of atomicity; each of them is a "photon" capable of exploding a satellite electron from an atom. The X rays are travelling at 186,000 miles a second (the speed of light) and the electrons, as already stated, at 10,000 miles a second. Most of the atoms are hydrogen atoms—or rather, since they have lost their satellite electrons, they are simply hydrogen nuclei or protons; these are travelling at 300 miles a second. Here and there we find heavier atoms, such as iron, lumbering along at 40 miles a second. Now you know the speeds and the state of congestion of the road; and I will leave you to imagine the collisions. It is not surprising that the atoms are found with their garb of electrons somewhat torn or even stripped naked.

These motions are those already referred to as distending the sun. We assign internal temperatures such that the corresponding motions are just sufficient to keep the sun distended to its observed volume. The calculation involves many technicalities which need not be mentioned here; but one datum is essential, viz. the average weight of the freely moving particles. The higher the average weight, the higher is the deduced temperature. The results are rather sensitive

* In this subject the opportunity of giving a number correct to 1 per cent. occurs so rarely that I have fallen to the temptation.

to this, so that unless the average weight is known fairly accurately the computed internal temperature, and more especially the computed luminosity, may be badly out. But how can we decide the average weight of the particles, being, as we are, ignorant of the chemical composition of the material in the interior? Here the ionisation of the atoms plays a very important part. We can best show this by a table for a representative selection of elements.

Element	No. of satellite electrons	Weight of atom	Average weight of particle
Lithium	3	7	1·75
Oxygen	8	16	1·78
Calcium	20	40	1·91
Iron	26	56	2·07
Silver	47	108	2·25
Gold	79	197	2·46

For example, if the sun were made entirely of oxygen and there were no ionisation, the average weight of the particles would be the atomic weight of oxygen (16); but ionisation splits each atom into 9 particles—8 electrons and a nucleus—and the average weight is therefore 16/9 or 1·78. The important feature is the steadiness of the average weight given in the last column of the table. It does not matter what element we choose from lithium onwards, or what mixture of elements; the average weight is always in the neighbourhood of 2. How different would it have been without ionisation! We should then have had a possible range from 7 to 197 instead of a possible range from 1·7 to 2·5; and we could not have made much progress without knowing definitely the composition of the star.

It would almost seem that Nature has taken a special interest in smoothing away our difficulties, for the small progression in the last column of the table is actually beneficial.

If by any chance the sun is made of gold its internal temperature is substantially higher than if it is made of oxygen (owing to the difference of 1·78 and 2·46). But we are not so much interested in the precise value of the internal temperature as in the flow of radiation through the star which results from it. Mass for mass, gold offers more obstruction to the passage of X rays than oxygen does, and this just about counteracts the effect of the higher temperature. In short, when we calculate the brightness of a star from its mass and radius we obtain practically the same result whether the material be oxygen or gold or any other element within the limits of the foregoing table.

But hydrogen is an exception. The hydrogen atom of weight 1 is broken into two particles, a proton and an electron, so that the average weight is $\frac{1}{2}$. This is too large a deviation from the normal value 2 to be ignored. If a large proportion of the material is hydrogen the internal temperature is substantially lowered and the calculated brightness is reduced very considerably. Broadly speaking we need distinguish only two kinds of matter inside a star, namely hydrogen and not-hydrogen.

I think that the one important change in the last seven years in the theory of the stellar interior is the recognition that hydrogen is very abundant.* You will find, for example, that in my book *Stars and Atoms* (1927) the conclusions are given subject to the reservation that there is not an excessive proportion of hydrogen (pp. 22, 36). We now believe that this proviso is not fulfilled. Ten years ago it was known that (on the usual assumption that the material was not-hydrogen) the calculated luminosities of the stars came out systematically too bright, and that this discrepancy could be cured by admitting sufficient abundance of hydrogen. Perhaps it will

---

* B. Strömgren, *Zeits. für Astrophysik*, vol. 4, p. 118 (March 1932); A. S. Eddington, *Monthly Notices of the R.A.S.*, vol. 92, p. 471 (April 1932).

be thought that the hydrogen explanation of the discrepancy might have been adopted then. But soon afterwards atomic physics was in the throes of a revolution, the older theory being replaced by wave mechanics; and until the laws of absorption of X rays were re-investigated on the new theory it was uncertain whether the discordance might not originate there. Gradually all loopholes seem to have been closed up, and we are apparently driven to adopt the hydrogen explanation. Simultaneously hydrogen has been found to be excessively abundant at the outside of a star, according to the interpretation of the spectroscopic observations.

There was another necessary step in the argument. Hydrogen, as the lightest of the elements, might be expected to diffuse to the outer part of a star; if so, it would not play the part we want it to play in lowering the temperature of the deep interior. Until this objection was met, it could not be claimed that abundance of hydrogen would cure the discrepancy between the calculated and observed luminosities. We now find that the slow diffusion of hydrogen to the surface is counteracted by a stirring of the material. The early theory of convection currents in the interior has been abandoned; but it is found that the rotation of the star must give rise to an up and down circulation which, although exceedingly slow, stirs the material faster than the hydrogen can work its way to the surface. The star may therefore be assumed to be of uniform composition throughout almost the whole interior.*

It seems possible now to make a reasonably trustworthy determination of the amount of hydrogen in a star if we know its mass and luminosity and, very roughly, its radius. We find that there is only one way in which such a mass

---

* The argument is not intended to apply to the outermost layers where the conditions are very different from those typical of the interior. Accordingly we must not assume that the uniform internal composition is the same as that found by spectroscopic examination of the surface.

could give the observed amount of light, namely it must consist of 30 per cent. hydrogen and 70 per cent. not-hydrogen, or whatever the calculated proportion may be. This refers to the composition of the interior—not to any visible region—and it is rather remarkable that there should

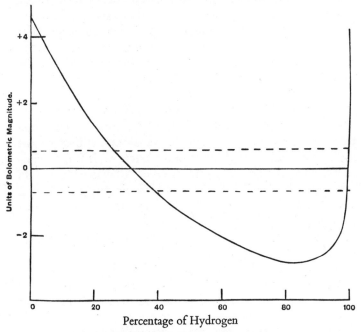

Theoretical Constitution of the Sun

be a way of discovering facts about the chemical constitution of such inaccessible matter.

The method of determination is illustrated by the diagram. The curve represents the calculated brightness of the sun (strictly the heat-brightness) corresponding to different assumed percentages (by weight) of hydrogen. The full horizontal line represents the observed brightness. The crossing-points, where calculation agrees with observation, accordingly give the possible percentages of hydrogen in the sun. There are two crossing-points, one at 33 per cent. and the other at 99·5 per cent. The first value seems more probable, especially when the corresponding results for other stars are taken into consideration; but there are some astronomers who advocate the other solution which exhibits the stars as globes of hydrogen with only a trace of the other elements. The second solution is ruled out if the heat of the stars is maintained by the transmutation of hydrogen (p. 167), for the sun would have consumed more than 0·5 per cent. of the hydrogen during its past history.

The diagram also shows two broken horizontal lines. These exhibit the uncertainty in the calculation which comes from our ignorance of how the subatomic source of the sun's heat is distributed through the interior. If it is all concentrated at the centre, we must take the intersections with the upper broken line; if the subatomic energy is being liberated evenly all over the sun without regard to temperature, we must take the lower broken line. It is unlikely to approach either of these extremes, so that the uncertainty from this cause is not serious.

A proportion of $\frac{1}{3}$ hydrogen and $\frac{2}{3}$ not-hydrogen* seems to fit most of the stars that have been investigated. There is some indication that the very massive stars have still more hydrogen. The difficulty is to find enough stars with well-determined masses to test such questions satisfactorily. One thing seems clear: stars of the same mass contain a remarkably constant proportion of hydrogen. It is hard to understand this constancy.

We cannot have it both ways; and since we use the com-

* Owing to the lightness of hydrogen the proportion by weight scarcely gives a fair idea of its abundance. The proportion, $\frac{1}{3}$ hydrogen and $\frac{2}{3}$ iron, would mean that there are 28 hydrogen atoms for 1 iron atom.

parison of the observed and calculated values of the luminosity to determine the (otherwise unknown) proportion of hydrogen, we cannot claim that the agreement furnishes any exact confirmation of the theory. Nevertheless there is a valuable check. It was by no means certain beforehand that any proportion of hydrogen would satisfy the observations. It will be seen from the diagram that if we do not know the proportion of hydrogen we can calculate a *minimum* brightness of the sun, corresponding to the lowest point of the curve. It is certainly a check that the observed brightness turns out to be above and not below the minimum brightness predicted by theory, and it is not so much above the minimum as to render the test a trivial one. The same test is satisfied by other stars covering a wide range of brightness, mass and density. When it is remembered that the minimum luminosity is calculated for the stars without knowing their chemical composition, without knowing how their heat is maintained, with no data except the mass and a rough knowledge of the radius, and using only the properties of matter determined under totally different conditions in the laboratory, it is very satisfactory to find that the actual luminosities of the stars run regularly a magnitude or two above the minimum calculated for them.

Fate has been rather unkind. After ten years doubt we seemed to have settled satisfactorily with the troublesome element hydrogen. And just then the physicists must needs discover a new element, neutron, of an equally upsetting kind. In 1924 I had to make the reservation "provided that the stars do not contain an excessive proportion of hydrogen". It seems that in 1934 I must make the reservation "provided that the stars do not contain a significant amount of neutron". The trouble is that neutron would make the material a very good conductor of heat. With any other material the leakage of heat through the star by conduction is negligible compared with the leakage by radiation. But it is said that 5 per cent.

of neutron would so greatly increase the conductivity that the whole of the observed outflow of heat would be attributable to conduction. I doubt whether we yet know enough about neutron to justify this estimate; but we must certainly keep an eye on the newcomer.

All the time our difficulty has been to construct a star which shall sufficiently hold in its internal heat, i.e. shall not be too bright. We have found it necessary to make it largely of hydrogen in order to keep the temperature low. I have busied myself with damming back the aetherial heat, but my efforts are of no avail if meanwhile neutron sneaks in and lets out the material heat.

It seems rather probable that there is a saving circumstance. The neutrons, or atoms of neutron, easily enter atomic nuclei; and presumably any neutrons evolved in a star will have only a brief free existence. We should expect them to be quickly snapped up by the atomic nuclei present, this being one of the processes of transmutation of the elements. Thus it may be hoped that the star will be kept clear of free neutrons, and that this threat to our conclusions will be countered.

IV

Referring once more to the title of Lane's paper, we notice the phrase "under the hypothesis of a gaseous mass". Lane knew quite well that the mean density of the sun is $1\cdot 4$ times that of water; so that it was rather an astonishing proposition to treat the material as though it were a gas. The practical astronomer could scarcely be blamed if he paid scant attention to a theory which took such liberties with the plain facts of the problem. Long after Lane's time it was discovered by Russell and Hertzsprung that there is a class of stars, called "giant stars", to which the theory can safely be applied because they have low densities like an ordinary gas. Capella, for example, has a mean density about equal to that of the air around us. Betelgeuse and Antares are more extreme examples of rarefied stars. By terrestrial standards we should describe Betelgeuse as "a moderately good vacuum". If our sun were distended to the dimensions of Betelgeuse it would envelop the earth's orbit, and we should be inside it.

It came to be accepted that Lane's theory applied only to the giant stars; and the other great division, the "dwarf stars" whose densities are comparable with those of terrestrial solids and liquids, was deemed to be outside its scope. Similarly the more extended investigations that I have been describing, including the theoretical calculation of the luminosity, assume that the star consists of perfect gas, and therefore the results were expected to apply only to giant stars. For such stars the luminosity depends on the mass, radius, and proportion of hydrogen; but the radius is comparatively unimportant, since any change of radius within reasonable limits is found to make very little difference to the result. Since the consideration of the hydrogen content is a later refinement, the calculation presented itself primarily as a theoretical relation between the mass and the luminosity of a star.

The mass-luminosity relation was calculated in 1924 and found to be satisfactorily obeyed by the giant stars. *But the agreement went too far.* Practically every star agreed with the formula—the rarefied stars for which it was intended and the dense stars for which it was not intended.

Consider, for example, the sun. If we make the rather strange assumption that its material is compressible like a perfect gas, its luminosity should be that given by the formula. But if its material is less compressible, it will hold its own against the pressure without requiring so much heat to keep it distended. The internal temperature will therefore be less than supposed, and not so much heat will leak out. Incompressible stars should therefore have a luminosity less than that given by the mass-luminosity relation. Conversely, if we find that a star obeys the mass-luminosity relation, that is evidence that its material is compressible like a perfect gas.

The sun and other dense stars obey the formula calculated for a perfect gas. The plain conclusion is that they are composed of perfect gas. Something in the condition of stellar matter—the extreme temperature or pressure—must have made it possible for material as dense as water or as iron to yield to pressure in the same way that an ordinary gas yields.

The explanation was not difficult to find. Why is it that we can compress air but we cannot appreciably compress water? It is because in air the ultimate particles—the molecules—are wide apart with lots of empty space between them. So when we compress air we are merely squeezing out this emptiness. But in water the molecules are practically in contact and there is no emptiness to be squeezed out. In all compressible substances the limit of compression is when the molecules begin to jam together. If a gas is compressed more and more, there comes a time when nearly all the empty space has been eliminated and the particles begin to jam; it then loses its characteristic compressibility and is said to have become "imperfect". By that time its density is more or less that of an ordinary solid and liquid. This refers to our terrestrial experience. But in a star the bulky terrestrial atoms and molecules no longer exist; most of the satellite electrons are torn away by ionisation. The lighter atoms are reduced to a bare nucleus of almost infinitesimal size; the heavier atoms retain a few of the closest electrons forming a structure perhaps $\frac{1}{100}$ of the diameter of a complete atom. So at the density of water or of the sun, when the complete atoms (if they existed) would be jammed in contact, there is still plenty of room between these tiny structures; and jamming will not occur until the matter is compressed to a density of the order 100,000 times greater. The material of the sun is therefore very far from its limit of compressibility, and it is right that we should find it behaving as a perfect gas.

THE CONSTITUTION OF THE STARS

Before this explanation could be accepted it was necessary to examine the effect of the electric charges of the ionised atoms. The effect is found to be small and actually tends to make the material slightly more compressible than a perfect gas.* This sounds paradoxical, for you would think that it would be more difficult to squeeze together ions which repel one another than complete atoms which are electrically neutral; but it must be remembered that the electrons which have been torn from their orbits are still present in the material, and they are able to compensate the nuclear charges as effectively as if they were bound.

We are therefore led to the conclusion that in stellar conditions the limit of compressibility will not be reached until the matter is perhaps 10,000 times denser than anything known on the earth. The most effective confirmation of the theory would be to find such dense matter actually existing. It happened that we knew where to look for it. In certain stars—three examples were known at the time—the usual method of determining the density appeared to fail for some unexplained reason. For the Companion of Sirius the method gave the absurd density of 60 kilograms to the cubic centimetre, or about a ton to the cubic inch. But if our theory is right, such a density is not necessarily absurd, and the method may not have failed after all. Accordingly W. S. Adams at the Mount Wilson Observatory undertook to check the deduced dimensions of the star, employing a method based

---

* This is partly a question of definition of "perfection". When an ionised gas is compressed at constant temperature, the ionisation diminishes. Neglecting the electrical forces, the reduced ionisation makes the pressure somewhat less than it would have been (at the same density) if the constitution of the gas had remained unchanged. Therefore, in comparison with an ideal gas of fixed constitution, we find that an ionised gas should be rather more compressible; the same increase of density corresponds to a smaller increase of pressure. The electrical forces *diminish* this super-compressibility, but they still leave the material slightly more compressible than an ideal gas of fixed constitution.

on Einstein's theory of gravitation. His results, which verified the high density, have since been confirmed at the Lick Observatory. It is generally accepted that the Companion of Sirius is an example of a star in which the material has an average density 2000 times greater than platinum. A match box filled with this material would require a derrick to lift it, since it would weigh about a ton. The dense material lies well below the surface at a depth where there is sufficient superincumbent matter to supply the necessary pressure. There is nothing abnormal about the layers which we actually see.

These super-dense stars are known as *white dwarfs*. They are probably very abundant in space; but since they have low luminosity we can only discover those that are in our immediate neighbourhood. Only three are definitely recognised, but several more stars have been assigned to this class on more or less probable evidence. There is also fairly strong ground for believing that the nuclei of planetary nebulae are white dwarfs. In these stars the matter is too near the limit of compression to be treated as a perfect gas, and they do not follow the mass-luminosity relation. As they form an exceptional class of great rarity from the point of view of the practical astronomer, we have not ordinarily the white dwarf condition in mind when speaking about the stars in general; and it must be understood that statements in which I attempt to convey the leading features of stellar constitution will not always apply to the white dwarfs.

V

It happened that just about the time that super-dense matter was discovered in the stars, an important development of wave mechanics was turning the thoughts of theoretical physicists in the same direction. R. H. Fowler was the first to recognise that the white dwarf stars provided a field of application for the "new statistics" which, according to wave mechanics, replaces the classical statistics of the ordinary theory of gases when the particles become crowded together. His treatment of the dense matter in white dwarf stars has been developed and extended by many subsequent writers. The theory depends primarily on the famous law in quantum theory called Pauli's Exclusion Principle. In its more special form of application it asserts that two electrons in an atom cannot occupy the same orbit (p. 35). More generally it requires that there shall always be a certain minimum of distinction between one electron and another. If the distinction had reference to position only, we could divide space up into equal unit cells and express the principle by saying that two electrons cannot be in the same cell. But the distinction also takes account of energy and momentum. That provides, as it were, another dimension in which distinction is possible if the distinction in position is insufficient. For a rough picture we can imagine the positional cells to form the ground floor of a sky-scraper. Two electrons cannot occupy the same ground cell; but one of them can occupy the corresponding cell on an upper floor if it has the energy corresponding to that elevation.

We imagine then the electrons to be living in a sky-scraper whose ground plan corresponds to space. The building is divided into rooms of uniform size, and a County Council regulation against overcrowding provides that two electrons may not occupy the same room. The electrons are moving and therefore continually changing their rooms. To mount to a higher floor the electron requires additional energy which it must absorb from radiation present in the material.* If it descends to a lower floor it emits energy. In a cold body at the absolute zero of temperature there is no radiation present; consequently the electrons may come down but they

* Disregarding mere exchanges of energy between the electrons by which one mounts up at the expense of another going down.

cannot mount up. In course of time they will all come down to the ground floor, *provided there are enough rooms there*.

The novel feature of very high density is that there may not be enough rooms on the ground floor, so that some of the electrons have to remain in the upper stories for lack of room below. In the more familiar low-density conditions there may be electrons on the upper floors; but this is mere exuberance of spirits—the result of a plentiful supply of energy. When the congestion on the ground floor begins, the pressure needed to compress the material is greatly increased, because it has not merely to pack the particles tighter but to lift them up to a floor where there is room for them.

A peculiarity of matter in this congested or, as it is generally called, *degenerate* condition is that, although it contains a great deal of what we should naturally call heat-energy, it is quite cold. The electrons relegated to the upper stories have energy; we picture them as travelling with great speed. But, unlike electrons similarly energised in uncongested conditions, they cannot spend their energy; it has, as it were, to be kept on deposit. Degenerate matter has thus a large latent heat, which is not available for radiation and does not take part in temperature exchanges. The latent energy can only be made available by allowing the matter to expand and become non-degenerate.

We cannot cover all the ramifications of the theory by an artificial picture of this kind: but the conception of a degenerate state of matter in which all the lower energy levels are filled up, and any additional particles forced in by compression have to be endowed with sufficient energy to occupy a high energy level, enters largely into the mathematical theory of dense stars.

The question arises, Is a high temperature necessary for attaining the white dwarf condition of matter? Supposing that we could apply sufficient pressure, would it be possible to crush cold terrestrial matter to a thousand times the density of platinum, or would it be necessary first to smash the atoms thoroughly by heating it up to 10,000,000°? It is now clear that pressure alone would suffice. The fragile shell of satellite electrons, which can be broken by the attacks of X rays or the fierce collisions in the interior of a star, can also break by simply giving way under the strain of pressure. Perhaps the strangest thing is that the compressibility of all kinds of matter—whether its density be that of a gas, of a terrestrial solid, or of the Companion of Sirius—is, apart from certain trivial aberrations, found to be much the same. There are two ways of reckoning the compressibility of material, according as the heat generated by the compression is or is not allowed to escape. We find the closest similarity if we adopt the second (adiabatic) reckoning. In a monatomic gas, e.g. helium, a 32-fold increase of pressure gives an 8-fold increase of density, if the heat of compression is retained in the gas. It is calculated that the dense matter in the Companion of Sirius is at least as compressible as this.

Why do terrestrial solids and liquids stand aside from the general rule that matter has a compressibility of the same order as that exhibited by helium? That is the trivial aberration referred to above. In the long run dense matter is not less compressible than rarefied matter, only its compression proceeds more jerkily. The apparent incompressibility of terrestrial solids and liquids is due to the fact that the ridiculously small pressures available to man are insufficient to get over the first jerk.

CHAPTER VIII

SUBATOMIC ENERGY

These people are under continual disquietudes, never enjoying a minute's peace of mind; and their disturbances proceed from causes which very little affect the rest of mortals. Their apprehensions arise from several changes they dread in the celestial bodies. For instance...that the sun, daily spending its rays without any nutriment to supply them, will at last be wholly consumed and annihilated; which must be attended with the destruction of this earth, and of all the planets that receive their light from it.

SWIFT, *Gulliver's Travels: A Voyage to Laputa*.

ARTIFICIAL transmutation of the elements was first accomplished by Cockcroft and Walton in 1932*. Up to that time our knowledge of the conditions of release of subatomic energy was derived almost wholly from astrophysical researches. In due time the data now being found in the laboratory will be of the utmost value to astronomy; we are on the threshold of big developments in the theory of stellar evolution and other problems depending on a knowledge of the source of a star's heat. But in this discussion I do not want to give too much prominence to our first hasty reflections on the new situation. We must wait until the present riot of experiment wears itself out a little. I would rather show the progress that astronomy has been able to make with the problem by its own resources, reserving until the end of the chapter the question how far the results are supported by the new laboratory discoveries. It would be premature to claim that the astronomical conclusions have

* Transmutation is produced by bombarding the nuclei with high-speed particles. By *artificial* transmutation we mean that the shower of bombarding particles is produced artificially. Some years earlier Rutherford had produced transmutations semi-artificially by using the high-speed particles emitted from radio-active substances.

SUBATOMIC ENERGY

been definitely confirmed; but they appear to be in keeping with the present trend of physics, and the opposition which they long encountered has died down.

Accordingly Sections I–III represent the outlook towards the end of 1932.* This enables us to introduce in Section IV the new experimental knowledge of the transformations of the atomic nucleus as entirely independent evidence bearing on the same questions. In so far as it is found (now or later) to lead to the same conclusions, it is a welcome corroboration of the general ideas and methods used in the study of stellar constitution.

I

I am going to tantalise you with a vision of vast supplies of energy surpassing the wildest desires of the engineer—resources so illimitable that the idea of fuel economy might be put out of mind. We have not to travel far to find this land of El Dorado, this paradise of power; the energy to which I am referring exists abundantly in everything that we see and handle. Only it is so securely locked away that, for all the good it can do us, it might as well be in the remotest star—unless we can find the key to the lock. We know very well that the cupboard is locked, but we are drawn irresistibly to peep through the keyhole like boys who know where the jam is kept.

We build a great generating station of, say, 100,000 kilowatts capacity, and surround it with wharves and sidings where load after load of fuel is brought to feed the monster. My vision is that some day these fuel arrangements will no

---

* In order the better to recapture the ideas of the time, I have followed in these sections as closely as practicable the text of a lecture given to the World Power Conference at Berlin in 1930, omitting or condensing those parts which would duplicate explanations given elsewhere in this book, and introducing only the minor modifications necessary to bring the astronomical statements up to date.

longer be needed; instead of pampering the appetite of the engine with delicacies like coal and oil, we shall induce it to work on a plain diet of subatomic energy. If that day ever arrives, the barges, the trucks, the cranes will disappear, and the year's supply of fuel for the power station will be carried in in a tea-cup, namely, 30 grams of water—or 30 grams of anything else that is handy.

I have called it a vision; but to the astronomer it means much more than an extravagant flight of theory. We look up at the sky and our telescopes show a thousand million stars. Everyone of these is a celestial furnace which apparently defies the law that limits our terrestrial undertakings—that if you do not continually replenish your furnace it will die out. Geological, physical, biological evidence seems to make it certain that the sun has warmed the earth for more than a thousand million years; but the calculation first made by Kelvin still stands incontrovertible that the sun's heat cannot have been maintained for more than twenty million years unless it is being fed from some secret store of energy of a kind unknown in his day. By all ordinary rules the sidereal universe which we see blazing with light should have long since been cold and dead. None of the sources of power utilised by our present civilisation could have kept it alive for more than a small fraction of the time it is known to have existed. It seems then quite plain that the "cup of water" method of maintenance is actually in operation in the stars, or that there is some partial adaptation of it. To the engineer the prolific liberation of subatomic energy is a Utopian dream; to the physicist it is a pleasant speculation; but to the astronomer it is just a common well-recognised phenomenon which it is his business to investigate.

As astronomers we have not merely to acknowledge the existence of sources of subatomic energy; we have to study observationally the laws of its release—to examine how the rate of liberation of subatomic energy varies with the temperature, the density, or the age of the matter concerned. We must also investigate how the supply is kept under control, so that a star fed with heat in this way is not blown to pieces or thrown into violent oscillation. A few general laws have been found in this way. It is true that they are only disconnected fragments of a complete scheme. But I have to insist that the study of subatomic energy is something imposed on us in the ordinary course of astronomical research, without which we cannot form any useful conclusions as to the evolution and general functioning of the stars. Like many lines of research in course of active development it is still in an untidy and unsatisfactory state.

Whilst insisting that it is a practical subject for the astronomer, I do not suggest that for the engineer it can be more than a dream for idle moments. I can see no escape from the conclusion that subatomic energy is the main fuel consumed in the celestial furnaces; but it would be wrong to raise illusive hopes that the astronomer may, like Prometheus, steal fire from heaven and make it available to men. Emerson's exhortation "Hitch your wagon to a star" is not to be followed literally by our transport authorities.

I have referred to the practical utilisation of subatomic energy as an illusive hope which it would be wrong to encourage; but in the present state of the world it is rather a threat which it would be a grave responsibility to disparage altogether. It cannot be denied that for a society which has to create scarcity to save its members from starvation, to whom abundance spells disaster, and to whom unlimited energy means unlimited power for war and destruction, there is an ominous cloud in the distance though at present it be no bigger than a man's hand.

II

Before jumping to the conclusion that the stars are utilising subatomic energy, there is a preliminary point to settle. Is it not possible that a star may be picking up from outside the energy necessary to maintain its radiation? Some have suggested that the sun is kept hot by meteors falling into it; others that it collects cosmic rays or still more subtle forms of energy traversing space. In short the question is, Does the star live on extraneous power like a windmill, or does it contain its energy stored inside it like an accumulator battery? I think that all theories which postulate an outside source can be dismissed, because they misconceive the nature of the problem. It is the temperature of some millions of degrees in the central regions that has to be maintained, and this requires the generation of heat in the deep interior. Meteors and cosmic rays provide only for keeping the *surface* hot. That is no use. For the maintenance of the sun's surface at 6000° would not stop the energy flowing out from the intensely hot interior, and the whole interior would presently cool down to the same temperature as the surface. We have seen that the internal heat is necessary in order to keep the sun distended to its observed volume. The problem of maintaining the sun's radiation is thus merged in the larger problem of maintaining its volume and other characteristics. We can only keep the interior of the star at a temperature of the order 10,000,000° by providing a source of energy in the deep interior. It appears then that a star contains within it the fuel that has to last it for the whole of its life.

The total store of energy contained inside the sun is easily calculated. Einstein has shown that there is an exact equivalence of mass and energy, such that 1 gram of mass represents 9.10^{20} ergs of energy. (The number 9.10^{20} is the square of the velocity of light in C.G.S. units.) We have only to convert the sun's mass 2.10^{33} gm. into ergs by this factor; the result

$1 \cdot 8 \cdot 10^{54}$ ergs is the sun's whole stock of energy. We know by observation that the sun squanders $1 \cdot 2 \cdot 10^{41}$ ergs every year by radiating heat and light into space; thus the whole stock amounts to just 15 billion ($1 \cdot 5 \cdot 10^{13}$) years' supply. By passing from the ordinarily known sources of energy to this store of subatomic energy we extend the possible life of the sun about a million-fold.

This does not mean that the sun will last 15 billion years and then go out; it is not quite so simple as that. The energy which we are now considering is energy of constitution of matter; and, of course, if you remove energy which is essential to its constitution the matter can no longer exist; it, so to speak, comes unstuck. And so as the stock of energy of the sun disappears little by little, the matter or mass of the sun disappears little by little. By the mass-luminosity relation (p. 153) the lower mass involves a lower rate of radiation. So we must allow for the fact that the sun will become less spendthrift in its old age; and its life as a waning star can be prolonged much beyond 15 billion years.

Similarly we can show that the sun cannot be more than 5 billion ($5 \cdot 10^{12}$) years old. Large masses radiate very strongly; and however large a mass it started with, the sun would have radiated itself down to its present mass within 5 billion years. I have never heard of any theory which required a longer past for the sun than that; but if anyone should propose a greater duration I think that astronomers would be justified in opposing it emphatically. It is more likely that we shall have to be content with a past duration of the sun very much less than this maximum estimate.

We have here been assuming a very drastic process of liberating subatomic energy, involving the complete disappearance of matter into radiation. For the matter to disappear it must be supposed that the protons and electrons of which it is composed have the power of mutually destroying one another. The proton carries a unit positive charge and the electron a unit negative charge, and it may be that under certain circumstances two such opposite particles can coalesce and cancel out. The idea is that when a proton and electron run together and neutralise each other, nothing is left but a splash in the aether representing the energy of constitution which is now set free. The splash spreads out as an electromagnetic wave, which is scattered and absorbed until it is converted into the ordinary heat of a star. The process is not difficult to imagine, but it is open to doubt whether it actually occurs in Nature. Apart from an indirect, and now very unlikely, inference from the phenomena of cosmic rays, there has been not the slightest observational evidence of its occurrence. Nor can it be said that it is a theoretical necessity that it should occur. It is just a conjecture. On the other hand I am not sure that it is more speculative to suppose that protons and electrons can end their existence in this way than to adopt the contrary view which supposes them to be immortal.*

There is a less drastic alternative. It is possible for matter to liberate some of the energy contained in it without going to the length of complete suicide. This alternative process is transmutation of the elements. By rearrangement of the protons and electrons in atomic nuclei a quite considerable amount of energy can be furnished. The most familiar example of such transmutation is in radio-activity. But none of the spontaneous radio-active transformations—uranium into radium, radium into lead, etc.—yields anything like enough energy for our purpose. Moreover it seems almost certain that a star is a place where radio-active elements are being synthesised, not where they break down. If the energy of radio-activity is of any account at all it must be reckoned a source of loss rather than of gain to the star; because presumably the transmutation is there proceeding the opposite way from that on the earth.

* See, however, p. 181.

The transmutation which might furnish sufficient energy to maintain the heat of the stars is the building up of complex elements out of hydrogen, more especially the formation of helium out of hydrogen. A hydrogen atom consists of one proton and one electron; a helium atom consists of four protons and four electrons, the four protons and two of the electrons being cemented together to form the helium nucleus. The material of a helium atom is thus precisely the material of four hydrogen atoms. But although the material is the same the mass is not quite the same; the helium is lighter by about 1 part in 140. By Einstein's law of the equivalence of mass and energy, this mass-defect is a measure of the energy that must be liberated when hydrogen is transmuted into helium.

You see then that there are two conceivable ways of getting energy out of four hydrogen atoms. They may disappear totally, each electron cancelling a proton; in that case the whole mass is lost, and the whole energy of constitution is set free to maintain the stellar furnace. Or they may rearrange themselves to form a helium atom; in that case $\frac{1}{140}$ of the mass is lost and $\frac{1}{140}$ of the whole energy is set free. Slightly more energy is set free if hydrogen is transmuted into a heavier element, e.g. oxygen, instead of into helium, but the advantage is trivial. Correspondingly the energy released in the transmutation of helium into oxygen is relatively insignificant. Recalling our previous classification of stellar matter as hydrogen and not-hydrogen, the only important source of energy is the transmutation of hydrogen into not-hydrogen; and this releases rather less than 1 per cent. of the whole energy. So if we decide to adopt the transmutation theory we must arrange to run the stellar furnaces on $\frac{1}{100}$ of the fuel available according to the annihilation theory. This cuts down the time-scale in the ratio $\frac{1}{100}$, and brings down the maximum life of the sun from birth to death to 150,000 million years. I daresay we can make that suffice. It is a more generous allowance than we need according to the results of Chapter x.

I have been speaking of the formation of helium and other elements out of hydrogen as though it were an established fact. It is true that no one has yet (1932) succeeded in performing such transmutations; but this objection seems scarcely relevant. It is an established fact that we find in Nature aggregations of protons and electrons in the particular formation which is called helium; and we are only applying the ordinary scientific outlook when we regard such a formation as brought about by the operation of physical law and not by an act of special creation. The evolution of our ordinary atoms out of their constituent electric charges must have occurred at some time and place. What place is more appropriate than the interior of a star, where the energy released by the process would serve for the maintenance of a star's heat? Can you suggest a more likely site for Nature's workshop where she forges a diversity of material out of the primitive basis of positive and negative electric particles? I have often encountered critics who argue that the stars are not hot enough for this purpose. I once so far forgot myself as to tell the critic to go and find a hotter place.

I will try to explain why it makes a big difference to astronomy which of the two possible sources of subatomic energy is in operation. Suppose first we assume the less drastic hypothesis of transmutation. Then 1 per cent. at the most of the total store of energy in the star is available to maintain its heat. Our peep through the keyhole showed us 100 pots of jam on the shelves; but it has turned out that 99 of them are unfit for consumption. As soon as it begins to shine, the star starts using up the one consumable pot, losing the corresponding amount of mass. When it has radiated away 1 per cent. of its original mass, the supply is finished; the furnace must die out and the star become cold.

Thus the mass of the star remains constant to within about 1 per cent. during its whole history. Contrast this with the other hypothesis of annihilation of electrons and protons. All the mass is then consumable. The dying out of the furnace is postponed, and the star may live to radiate 50 per cent., 75 per cent., 90 per cent. of the mass it had to start with. Beginning as a heavy-weight star it will gradually change into a light-weight star. By the mass-luminosity relation its brightness will diminish as its mass diminishes. We shall have an evolution of small stars from large stars, of faint stars from bright stars. Many interesting astronomical results arise out of tracing the consequences of this evolution. But all this falls to the ground if we reject the annihilation hypothesis and admit only transmutation. There is then no appreciable change of mass, and small stars differ from large stars because they were born different. So until we can decide between the two hypotheses we are like children speculating whether ponies grow into horses or whether ponies and horses have always been different.

It suggests itself that we should try to make an observational test whether big stars turn into little stars. It may not be an infallible test, but it is a fairly direct test. Let us take all the brand-new stars we can find, and see what sort of mass they have. I think it is fair to assume that the most recently formed stars are those with low density; we believe that the stars have condensed out of nebulous material, so the first stage should be a huge diffuse globe—a star such as Betelgeuse or Antares. Taking a list of about 300 of the most diffuse stars and calculating their masses from their brightness by the mass-luminosity relation, we find that their mean mass is $3 \cdot 6$ times that of the sun, and nine-tenths of them are between $5\frac{1}{2}$ and $2\frac{1}{2}$ times the sun's mass. It appears therefore that the stars at birth seldom or never have a mass so low as that of the sun. On the other hand taking stars of all ages there are far more masses below that of the sun than above it. It would seem that these must have lost a great part of their original mass, having radiated it away in the course of billions of years as the annihilation theory suggests.

Unfortunately this is counterbalanced by evidence of another kind which is unfavourable. We sometimes find clusters of associated stars, which evidently have a common origin and must have been formed about the same time. The Pleiades is a well-known example. The theory requires that these coeval stars should be of nearly the same mass and brightness. For if the cluster is young, there has not been time to radiate the large original masses down to the sun's mass or lower. If the cluster is old, the original range of mass will have been lessened in the course of time; because the large stars radiate away their mass very quickly and so tend to catch up the smaller stars which radiate more slowly. But this is not at all in accordance with observation. In the Pleiades the stars range over at least 10 magnitudes—indicating a wide diversity of mass. We have to admit that, in the Pleiades at least, small stars are born small and not evolved out of big stars. Such exceptions make us very sceptical about the whole idea.

One could cite other considerations of a similar kind, some rather favourable to, others rather against, the annihilation hypothesis. It is all very inconclusive. In studying the stars as individuals there is, in spite of some difficulties, much that attracts us to the hypothesis and the extremely long timescale which results from it. But when we turn to consider systems of stars—clusters and galaxies—all the evidence indicates much less antiquity. It now seems very unlikely that we have to go back more than 10,000 million years. I am sorry to be so vacillating—in one argument putting the beginning of things as we know them billions of years ago, and a few pages later lopping off two or three noughts from that figure; but everything depends on which line of circumstantial evidence you trust.

We shall see in Chapter x that the rapid expansion of the universe points strongly to the shorter time-scale. Additional support is given by a study of the dynamics of our own galaxy. It can be shown that the rotation and distribution of stellar motions which we find in our galaxy is incompatible with a strictly steady state of the system; and it appears that change and dissipation must be fairly rapid.* Whereas a star considered by itself is something which, so far as we can tell, might have existed with very little change for untold ages, the vaster systems—clusters of associated stars, our own galaxy, and the whole super-system of the galaxies—are in much more of a hurry to get on with their evolution. They are not yet worn down to regularity, and bear the marks of comparatively recent origin. Comparatively recent, I may remind you, means in this connection something of the order 10,000 million years instead of the alternative suggestion of 10,000,000 million years.

III

Leaving the decision between the two possible sources of subatomic energy unsettled, we now consider the astronomical evidence as to the conditions which govern its release. On this side of the problem we seem to have quite definite information—if only it were not so incredible! Apparently if you want to tap a really large supply of energy you must heat matter up to a temperature of about $20,000,000°$ Centigrade. I will not guarantee that $20,000,000°$ is exactly the right figure; it may be $15,000,000°$ or a little less. But my point is that there is a temperature somewhere about this magnitude at which matter yields up its energy prolifically, whereas one or two million degrees below it the yield is practically *nil*. It is almost like a boiling point.

* This is discussed in my Halley Lecture, *The Rotation of the Galaxy* (Oxford, 1930).

The stars are now classified into three groups, the giants, the main series, and the white dwarfs. The giants are comparatively few in number and presumably represent an early and rather transient phase of development. The white dwarfs are probably numerous, but owing to their low luminosity very few are actually known to us. By far the majority of the stars that we investigate belong to the main series. The main series forms a continuous sequence extending from the brightest to the faintest stars known. It is found that, from the top to the bottom of the series, the central temperature (calculated by the methods explained in Chapter VII) remains practically constant at about 20,000,000°. At the top of the series we have very bright and massive stars radiating 10,000 times as much energy as the sun; to keep up this output they require a continual supply of released subatomic energy amounting to 1000 ergs per second per gram of material. Near the middle of the series we have the sun, which requires 2 ergs per gm. per sec. to maintain its output. At the bottom we have stars requiring ·01 erg per gm. per sec. But whether the amount required is 1000 or 2 or ·01 ergs per gm. per sec., the temperature has had to rise to 20,000,000° in order to set it free.

So long as the subatomic energy liberated in the interior is less than the amount of energy squandered in radiation the star must go on contracting; and if it is an ordinary star (not a white dwarf), its internal temperature will rise. The rise continues until the conditions become such that the necessary amount of subatomic energy is liberated. When this balance is reached the star remains practically steady for an enormously long period, and we should expect to find the majority of the stars in this state. On the annihilation hypothesis the mass gradually diminishes and the star travels slowly down the main series. On the transmutation hypothesis the star remains stationary on the line of the main series until its hydrogen is mostly used up; presumably it

SUBATOMIC ENERGY

then passes on to the white dwarf stage. In either case it is clear that a very rapid increase in the liberation of subatomic energy must set in at about 20,000,000°, since stars requiring widely different amounts find their balance at about this temperature.

Is this the key to the cupboard? Suppose we could manage to heat terrestrial matter up to 20,000,000°, should we extract its energy of constitution? I may remark in passing that if this is the method required, the chances of our making a commercial success of it are not very promising; we should waste a lot of power in maintaining the high temperature whilst the stream of subatomic energy dribbles out. But I scarcely think it can be the key. It must have some bearing on the problem; but a more general survey of the difficulties than I can give here convinces me that there is a great deal more that we shall have to understand before we can put these astronomical results in their right perspective. Twenty million degrees is perhaps not beyond attainment in our laboratories. At the Cavendish Laboratory Prof. Kapitza produces momentary magnetic fields in which the concentration of energy corresponds to about 1,000,000°. If he should be able to raise this to 20,000,000°—Well, I have said that I do not really expect the subatomic energy to come pouring out; but all the same I shall not go too near the laboratory when the experiment is tried.*

There is another condition of release which is of great importance in astronomy. It is necessary that the liberation

---

* This was written when we had no theoretical knowledge as to the cause of the critical temperature, and the possibility that it might be a genuine "boiling point" had to be reckoned with. I think there is no doubt that matter containing hydrogen, e.g. water, is a high explosive in the sense that the sudden generation of sufficiently high temperature would release subatomic energy so fast that the temperature would be maintained and spread through surrounding matter regeneratively; but it now seems clear that the regenerative temperature is considerably higher than 20,000,000°. (See Section IV.)

of subatomic energy should be stimulated by increase of temperature; otherwise it will not automatically adjust itself to keep the star steady for long periods of time, and subatomic energy will therefore fail to serve the purpose for which we have introduced it. But it must not increase too fast with the temperature, because that would have the effect of throwing the star into pulsation. It is very probable that some stars do pulsate, alternately swelling out and contracting in a period of a few days or hours; they form a class of variable stars called Cepheid Variables. At one phase of the pulsation the star's material is compressed and hotter than the average; at the opposite phase it is expanded and cooler. The subatomic supply of heat will be stimulated by the increased temperature at compression and reduced by the lower temperature at expansion. Now this is just the way in which the heat supply of an engine must be regulated in order that the engine may be set working; heat must be supplied to the cylinder at compression and removed at expansion. Thus the star becomes an automatic engine which can maintain its own pulsations, or even work up a large pulsation out of a very small initial disturbance. The puzzle is, not to explain the Cepheid Variables, but to explain why they are the exception and not the rule.

The pulsation will be attended by a certain amount of wastage; and the occurrence of the pulsation depends on whether the engine-effect that I have described is strong enough to make good the wastage. If the resistance is too great an engine will not start up. We must suppose that in the sun and in ordinary stars the engine is not strong enough to keep a pulsation going. That is one of the conditions to which we have to attend in formulating the laws of release of subatomic energy; they must not provide too powerful an engine. In other words the release must not be stimulated too rapidly by a rise of temperature above the normal temperature in the star. We can calculate roughly how rapid a rate of increase with temperature is permissible.

This puts us in a dilemma. By comparing the temperatures of the various stars on the main series, we have seen that the increase in the rate of liberation of subatomic energy from ·01 to 1000 ergs per gm. per sec. must occur within a range of temperature too small for us to detect with certainty, say two or three million degrees. This is much too rapid a rate of increase to satisfy the new condition that we have found.

Apparently the only way out of this difficulty is to suppose that the stimulating effect of an increase of temperature is *delayed*. There must be a time lag—anything from a few days to a thousand years—between the rise in temperature and the corresponding increased output of energy. That is to say, when the increase of temperature occurs there is no great immediate increase in the production of energy, but there is an increase in the production of an active kind of material which in due course (after some days or perhaps years) undergoes a spontaneous transformation which liberates subatomic energy. Such a time lag would smooth out the effect of the rapid changes of temperature in a pulsation; for it makes the rate of liberation of energy depend on the average temperature during the period of the lag. This will save the star from being thrown into pulsation. On the other hand it would make no difference to the permanent adjustment of the rate of liberation of subatomic energy to the rate of radiation of the star.

We can now sum up the astronomical evidence concerning the liberation of subatomic energy:

(1) There is abundant liberation of some form of subatomic energy at a comparatively low temperature of the order $20,000,000°$ or rather less.

(2) Unlike ordinary radio-activity it is affected by the physical conditions of the material, and the liberation increases very rapidly with the temperature.

(3) There is a time lag between the change of temperature and the corresponding change in the rate of liberation. This signifies that unstable material is formed which after some

days or years spontaneously breaks down; and it is in this subsequent break-down that the greater part of the energy is liberated.

(4) Evidence as to whether the source of the energy is transmutation of hydrogen or annihilation of protons and electrons is inconclusive; but the recent tendency is to favour the former with its accompanying short time-scale of evolution.

IV

Among physicists generally there was a great reluctance to accept the conclusions (1) and (2) in the foregoing summary. There were also astronomical critics. It was continually urged that subatomic processes could only be influenced by temperatures a thousand or a million times greater than those which we have found in the stellar interior. It is for this reason that I have called 20,000,000° a "comparatively low temperature". Throughout the last fifteen years there have been attempts to find a loophole for attributing a much higher temperature to the centre of a star, or alternatively to work the machinery of a star with an unadjustable (radio-active) source of energy unaffected by temperature and density. These have seemed to me to ignore one or more of the essential conditions of the problem, and to subordinate that branch of the subject—the mechanical and thermal equilibrium of the star—which depends on fairly well-known laws of physics to speculation on matters about which the physicist knew even less than the astronomer.

The recent achievement of artificial transmutation of the elements in the laboratory has brought about a revulsion of feeling, and the astronomically determined temperatures of the stars are no longer criticised as too low. The transmutation is accomplished by subjecting the atomic nuclei to bombardment by particles of various kinds—protons, electrons, neutrons, deutons, helium nuclei (α particles).

A certain proportion of these hit the nuclei and enter them. The particle may simply be retained, or its ingress may upset the equilibrium of the nucleus in such a way that some other kind of particle is expelled; in either case the constitution of the nucleus is changed and it becomes a different element. Here we are chiefly interested in the entry of protons (hydrogen nuclei) into the more complex nuclei, for we have seen that the astronomically significant liberation of energy (if any) comes from the transmutation of hydrogen. It is found that no great energy is required to enable a proton to penetrate a nucleus; the lowness of the energy seems to have come as a surprise to the experimenters. The progress of artificial transmutation in 1933 was made possible not by the use of unprecedentedly high voltages but by the great advance in the sensitivity of the methods of detecting transmutations.

The average energy of the particles (including the protons) near the centre of the sun is equal to that imparted to a proton when about 2500 volts are applied. There will always be a few protons with energies many times greater than the average—comparable therefore with the energy of the protons employed in artificial transmutation. We do not want the conditions to be such that the protons enter the nuclei very often, for the sun's supply of hydrogen has to last it for at least 10^{10} years. Until more detailed laboratory data are available it is impossible to make a precise comparison, but the general estimate is that at somewhere between 10,000,000° and 20,000,000° the protons (or hydrogen) would disappear into the nuclei quite fast enough to provide the energy used to maintain the sun's heat. The transmutation is very sensitive to an increase of voltage, and correspondingly to an increase of temperature; so that stars requiring widely different supplies of energy will find their equilibrium at temperatures within a rather small range. Thus the observational result which at first seemed so incredible is confirmed.

It should be noted that even if we prefer the hypothesis of annihilation of electrons and protons as the main source of a star's energy, we must not disregard the effects of transmutation of hydrogen. The transmutation of hydrogen will act as a buffer preventing the temperature from rising above 20,000,000°—so long as any appreciable amount of hydrogen remains in the star. For if the star contracts so as to raise its temperature, the protons will attack the atomic nuclei more frequently; more energy will be liberated, which will cause the star to expand again and the temperature to fall. Theories, advocated until recently, which attributed temperatures of thousands of millions or billions of degrees to the stars, are now quite out of the question—unless the stars are assumed to be almost devoid of hydrogen in their interior. At such temperatures matter containing hydrogen would be a high explosive.

There is as yet no direct confirmation of the time lag in the liberation of the energy (p. 175). On the other hand it is no longer a surprising conclusion; for in the bombardment of atomic nuclei with various particles (but, I think, not as yet with protons) it is often found that unstable nuclei are created which break down and give out energy after a few minutes or hours.

The new discoveries may perhaps have removed one of the difficulties in the conception of evolution of complex elements inside a star. Formerly we knew of nothing intermediate between a proton and a helium nucleus. Thus the first step in evolution appeared to be the gathering together of 4 protons and 2 electrons to form a helium nucleus. How these could assemble simultaneously at one spot baffled imagination. We could only comfort ourselves with the reflection that they obviously *had* managed to assemble, and that the interior of a star could scarcely be a less favourable place for the purpose than anywhere else. But now neutrons, deutons, and isotopes both of hydrogen and of helium of weight 3, have been discovered, all intermediate between a proton and a helium nucleus. Thus helium may be built up gradually by the same kind of steps that occur in the evolution of the higher elements.

An alternative possibility (suggested and developed by R. D'E. Atkinson in 1931) is that the helium is formed inside complex nuclei and then expelled. To take an ideally simple example, we can suppose that protons and electrons enter a complex nucleus one by one, where they arrange themselves as far as possible as α particles. Now and then the structure collapses and an α particle (helium nucleus) is expelled. This may happen over and over again in the same nucleus. Assuming that, if the helium nucleus is accounted for, there is no difficulty in its further transmutation into a more complex nucleus, the progeny of helium nuclei will in due time provide additional complex nuclei to carry on the work. A single helium atom might in this way be the ancestor of all the not-hydrogen in a star.

At one time it seemed that Cosmic Rays might have an important bearing on the problem of subatomic energy. Cosmic rays is the name given to a highly penetrating radiation (consisting either of electromagnetic waves or particles) which travels downwards through our atmosphere, apparently having come into it from outside. It has been a favourite hypothesis that they have their birth in subatomic processes occurring in the nebulae or cosmic clouds in our own and other galaxies; they have been variously attributed to the transmutation of hydrogen into particular elements or to the annihilation of electrons and protons. Attempts to identify the process originating them depend on our knowing the energy of an individual ray, and until recently this could only be inferred from measurements of the penetrating power. It now appears that the energy of the strongest rays was very much underestimated, and previous interpretations have had to be revised. When stopped by matter a cosmic ray sometimes produces a great shower of electrons and positrons*; these can be traced individually in a Wilson expansion chamber and their energies of projection measured and summed. The original energy of the ray must be not less than this total. It turns out to be very much greater than the energy of any individual subatomic process admitted by existing theory. The cosmic rays are still a great mystery; but in view of their excessive energy it now seems impossible to attribute to them a subatomic origin.

In dismissing cosmic rays from our subject we must dismiss with them certain ideas for which they were responsible. It was clear that they could not come from the hot interior of a star, because they could not pass through any considerable part of the thickness of the star. They had therefore to be attributed to diffuse matter through which they would have practically free passage. The observed intensity of the cosmic rays indicated that the comparatively cool diffuse matter of the universe must be liberating energy not much less abundantly than the stars themselves. Against the natural conclusion from stellar observation and theory that the liberation of subatomic energy depends on the rather high temperature in the interior of the stars, we had to set the apparent evidence of the cosmic rays that high temperature is by no means essential inasmuch as similar liberation occurs in nebulae. The latter evidence has proved untrustworthy, and there is now nothing to distract us from the stellar clues.

V

The discovery of the positron deals a blow to the annihilation hypothesis. We now know that the positron, not the proton, is the true enantiomorph of the electron. A positron and an electron *can* annihilate one another. The experimental evidence seems conclusive that twin electrons and positrons are created

* See Plate 1.

when radiation of sufficient energy falls on matter, and that after a brief existence the positron ends its life by mutual suicide with an electron. Of course, this does not prove that an electron cannot equally end its existence by cancelling a proton; but the hypothesis begins to look rather gratuitous.

The discovery of the neutron also makes a difference. One has the feeling that the combination of proton and electron in a neutron is the nearest they can go to cancelling one another. In a sense it is not far off cancellation, for the neutron is, as we have seen, "an isotope of nothing". A neutron is so elusive, and has so little interaction with the matter through which it passes, that it is hard to detect that there is anything there. Having discovered this form of intimate combination of a proton and an electron—a state of zero quantum number —we feel it unlikely that there is yet another kind of combination resulting in complete destruction.

To this I may perhaps add a personal view, based on the way in which the combined relativity and quantum theory is working out, that there are conditions which fix for all time the net number* of electrons and protons in the universe.

Although I have not ventured to go so far elsewhere in this book, I think the time has come to consider whether the hypothesis of annihilation of electrons and protons might not be allowed to lapse. I can perhaps suggest this the more freely because I think that as an astronomical hypothesis it first occurs in my own writings,† although the general idea was familiar enough to physicists at the time. As in the case of determinism, it is not a question of asserting definite disproof, but of realising that it is no more than a survival from a time when the state of our knowledge was different from that prevailing to-day. When the hypothesis was first

* Counting a positron as "minus an electron", and a negatron as "minus a proton".
† *Monthly Notices of the R.A.S.*, vol. 77, p. 611 (1917).

suggested no other adequate means of maintaining a star's energy was known. It was not until 1920 that Aston's accurate determination of the atomic weight of hydrogen revealed the large amount of energy to be obtained by the transmutation of hydrogen into not-hydrogen, and so provided a possible alternative. We have seen (p. 168) that a decision between the two alternatives was not to be undertaken lightly, owing to its profound effect on our views of stellar evolution; and indeed the annihilation hypothesis was at the time the more conservative, being less disturbing to the current theory. Since then the relative status of the two hypotheses has changed in the following ways:

(1) Transmutation is now a matter of practical knowledge and is studied in detail in the laboratory. It is known to occur in conditions corresponding to the temperature of the stars. We have in any case to take account of its effect on a star's supply of energy, whether or not it is the sole source. On the other hand there is no observational evidence of annihilation; the cosmic rays which were sometimes dubiously regarded as giving such evidence are now found to have a different origin.

(2) It now appears inevitable that we should accommodate ourselves to the shorter time-scale, and the main advantage of the annihilation hypothesis disappears. Accepting 10,000 million years as an upper limit to the age of the stars, the sun's heat would be maintained for this period by transmuting an amount of hydrogen equal to 10 per cent. of its mass. In this connection the discovery of the great abundance of hydrogen in the stars (p. 147) is a favourable point.

(3) From the theoretical point of view the cancelling of an electron and proton is not so natural a suggestion as it formerly appeared. Larmor's picture* of the creation of a positive and a negative particle by rotating the walls of a tube with respect to an inner core—with the possibility that

* *Aether and Matter*, Appendix E (1900).

the walls may ultimately slip back, annihilating the two particles—is now seen to refer to the electron and positron rather than to the electron and proton.

The present moment, when there is a rush of new discovery only half digested, is not the best time for making up our minds whether the hypothesis of annihilation is worth preserving. It will be apparent from many passages in this book that I have not yet taken the step of retiring it from my own thoughts. It is doubtless best to leave the question in abeyance for a year or two longer, but it has seemed well to call attention to its imminence.

CHAPTER IX

COSMIC CLOUDS AND NEBULAE

> When I behold, upon the night's starr'd face,
> Huge cloudy symbols of a high romance.
>
> KEATS, *Sonnet*.

I

I AM going to speak about a very rarefied cloud of gas which occupies all the space between the stars. First let me remind you of the vastness of this space and the extreme isolation of the stars from one another. The stars are small oases of matter in a desert of emptiness. For a traveller in this desert we may take a ray of light. His journey from one oasis to the next, say, from the nearest star to our sun, takes four years; he takes only eleven hours to cross the whole extent of the solar system; and then the journey is through empty desert again for six years or so. That is if the light ray were to zigzag from star to star; if it goes unheedingly on a straight course through the universe it will probably miss the oases altogether as a traveller in a desert would do.

But this space between the stars, which I have called a desert of emptiness, is not entirely empty. There are vestiges of matter everywhere. In some parts of the heavens we can actually see a rarefied cloud amidst the stars. Examples are shown in Plates 2 and 3. In one the nebulous matter is bright —wisps of glowing gas wreathed into a delicate lacework. In the other there is, besides bright matter, an impenetrable black cloud blotting out everything behind it. It is only in certain regions that we see it thus plainly, but the cosmic matter extends everywhere. The recognisable nebulae are condensations—places where the density is sometimes as

PLATE 2

Mount Wilson Observatory

GASEOUS NEBULA

N. G. C. 6992 in Cygnus

COSMIC CLOUDS AND NEBULAE

much as a thousand or ten thousand times the normal. I shall first speak of the normal regions, where accordingly the photographs give no indication of matter being present. The invisible gas filling these regions will be called the "cosmic cloud" or "interstellar cloud". We ourselves are probably in a normal region where the cloud has more or less its average density.

Until about ten years ago astronomers had no very satisfactory evidence of the existence of the cosmic cloud; nevertheless it has been a subject of discussion for forty years or more. Our former attitude towards it reminds me of the guest who objected to sleeping in the haunted room. "But I thought you did not believe in ghosts!" "I don't believe in ghosts, but I am afraid of them." Probably not many astronomers believed in the cosmic cloud, but some of them were afraid of it. Afraid, because, if such a feature of the stellar universe existed unheeded in our calculations, it might upset some of our most fundamental conclusions in astronomy. Having measured the apparent magnitude and distance of a star, we can calculate its true brightness—provided it may be assumed that we see it undimmed by intervening fog. Interesting conclusions may be drawn from a dynamical study of the motions of the stars—but it is assumed that the movements are not interfered with by a resisting medium. We calculate that in the course of time the masses of the stars must decrease by the loss of mass due to radiation—but what if at the same time the stars are acquiring more mass by sweeping up the cosmic cloud as they pass through it? The cosmic cloud was thus a bogey which threatened the security of many of our theories of the structure and mechanism of the stellar universe. And so there arose discussions and theories of the cosmic cloud and attempts to estimate its probable properties. This was not speculation; it was precaution. Now that the bogey has materialised it has lost its frightfulness; it turns out that the cosmic cloud is so sparse that it is not a very serious factor in the problems I have mentioned, though it is perhaps not always negligible.

I suppose it was in any case improbable that interstellar space would turn out to be entirely empty. Nature abhors a vacuum; and we must expect individual atoms to stray away from stars and nebulae and get lost in the vast regions of space, much as dust accumulates in an empty room. We generally suppose that the stars have condensed out of one primordial nebula comprising the whole galaxy, and we can calculate that the condensations would not entirely drain the matter from the regions between them. Thus we may expect to find the universe a bit dusty, either by accumulation or because it was not properly cleaned to begin with. It is true that a certain amount of sweeping goes on. The stars, like celestial housemaids, run hither and thither, and by their gravitation draw in the surrounding matter. But the sweepers are few compared with the volume to be swept, and we can calculate that by this process it will take at least 10,000 billion years to complete the celestial spring-cleaning.

II

I will come at once to the direct evidence for the existence of a cosmic cloud. It is well known that when light passes through a gas the atoms leave their characteristic mark upon it, so that when the light is analysed by passing it through a prism the spectrum shows a number of gaps or dark lines. These gaps, which represent the depredations of the atoms, indicate not only the chemical nature of the gas but how fast it is moving towards or away from us. For example, if we turn a spectroscope on to one edge of the sun we see the lines of a gas, e.g. iron vapour, in a position which indicates that the gas is coming towards us; turning it on to the other edge of the sun we see the same set of lines but they are now in a position which indicates that the gas is going away from us. One edge coming towards us and the other going away from us means that the sun is rotating—a fact already discovered by watching the sunspots which appear from time to time on its surface. But there are a number of lines in the sun's spectrum which do not show this effect of rotation; they are seen in the same position whether we look at the east or the west edge of the sun. Clearly they are not formed in the rotating atmosphere of the sun; they must have been imprinted on the light after it got clear of the sun altogether. We have discovered a stationary gas lying somewhere between the sun and our telescope. Moreover we have discovered its chemical composition; the stationary lines correspond to oxygen and nitrogen. A stationary medium consisting of oxygen and nitrogen—Why! Of course there *is* a stationary medium consisting of oxygen and nitrogen between the telescope and the sun. It is only our own atmosphere we have rediscovered.

The same method applied to the stars has, however, had more momentous results. The effect was first noticed by J. Hartmann in 1904 in δ Orionis, which is one of the three stars in Orion's belt. It is a double star, but most of the light comes from the brighter component, and the spectrum of the fainter component is not visible. We can follow the motion of the bright component in its orbit by observing the lines of its spectrum. For three days the bright component comes towards us and the dark lines are seen shifted towards the violet; then for three days it recedes and the dark lines are seen shifted towards the red. This applies to most of the dark lines. But there are two strong lines due to the element calcium, known as the H and K lines, which remain in the same position all the time. Evidently these have a different origin from the others. They are imprinted on the light after it has left the moving star, and indicate some medium containing calcium vapour which lies between the star and our telescope. It is not the earth's atmosphere this time, for that does not contain calcium vapour. And in any case by measuring the positions of the H and K lines we determine the motion of the calcium vapour, and find that it is not connected with the earth just as we have found that it is not connected with the star.

The only other "fixed line" that has been observed is the yellow D line of sodium. These lines of sodium and calcium, seen in the spectra of stars but evidently not belonging to the stars, have been found in the spectra of a great many stars. It seems a natural inference that the calcium and sodium form a cloud diffused through interstellar space, through which the light of the stars travels to reach us. This hypothesis was in fact proposed by Hartmann, but it was a long while before it became accepted. The objection was that only the very hottest types of stars (Types O and B) show the fixed lines. It was argued that if the lines were formed in a medium filling interstellar space all classes of stars ought to show them. We shall see later how this objection has been met. In the meantime it seemed that the high temperature of the stars must have something to do with the phenomenon, and therefore the calcium and sodium vapour must be comparatively close to the star. The common belief was that it formed an aureole enveloping the whole double star; the two component stars pursued their orbits within this envelope without disturbing it seriously. This could be put to the test. So far as periodic orbital motion is concerned the calcium-sodium envelope need not follow the star moving to and fro within it; but the average motion over a long time must be the same for both, otherwise the star and its envelope would separate. This test indeed had been thoroughly applied as early as 1909 by V. M. Slipher, who reached conclusions which accord with the modern results; but his work seems to have been overlooked.

In 1923 an investigation by J. S. Plaskett with the 72-inch reflector at the Dominion Observatory, British Columbia, removed all doubts on this point. Observing some forty stars which showed fixed lines, he found that there were considerable differences (sometimes very large differences) between the velocity of the star and the velocity of the calcium. Interpreted according to the foregoing view, the stars were leaving their haloes behind. An equally significant fact was that, whereas the stars had motions of their own, some large, some small, the calcium was always found to be nearly at rest in space. Not at rest relatively to the solar system, for the sun has an individual motion of its own; but relatively to the more significant standard "the mean of the stars", the calcium sampled in different parts of the sky was found to have little or no motion. This strongly suggests that it forms one continuous cloud.

This is the primary evidence which leads us to picture a cloud of matter filling the stellar system, comparatively quiescent, with the stars rushing about through the midst of it. Light sets out from a distant star on its journey towards us travelling 186,000 miles every second. On and on it goes, year after year, with sparsely strewn atoms in and around its track. Now and again a calcium or a sodium atom makes depredations. The light has to run the gauntlet for, say, 1000 years before it reaches the earth. It arrives depleted in those constituents which calcium and sodium atoms devour, showing therefore those gaps (dark lines) in its spectrum which have enabled astronomers to unravel the story.

The longer the light journey the greater the loss by depredation. Therefore the intensity—the blackness of the stationary calcium lines—should be a clue to the distance of the star. That was the next test to try. It was first shown to be fulfilled by Otto Struve. We scarcely expect this relation of intensity to distance to be very accurate because the cloud will not be uniform; the nebulae, for example, are places where it is strongly condensed. But smoothing out irregularities by taking the mean results for stars at different distances, the increase of intensity with distance is quite marked; moreover it increases according to the law which the theory of absorption would lead us to expect.

A particular example may perhaps be more impressive than general statistical confirmation. It was noticed in 1910 that the stars of high temperature in and around the constellation Perseus divide themselves into two groups according to their proper motions. In the foreground there is a group of stars, all moving across the sky in the same direction and apparently at the same rate, evidently forming an associated cluster. The apparent motion is large for this class of star, so that it is fairly certain that the cluster is relatively near. The remainder of the stars in the region show little or no apparent motion and form a distant background. This is a good opportunity for applying the test, because in observing foreground and background stars we are looking in the same direction through the cloud and are not so liable to be misled by irregularities of its density. It is found that the foreground and background stars can be distinguished at once by the intensity of the fixed calcium lines; these show up much more strongly in the background stars, owing to the greater thickness of cloud in the way.

A still more remarkable test has been applied by Plaskett and Pearce. It depends on the fact that the whole of our galaxy of stars is rotating about a centre far away from us in the direction of the constellation Ophiuchus. It is not rotating like a rigid body, but (as required by the law of gravitation) the outer parts revolve more slowly than the inner parts—as the outer planets in the solar system revolve more slowly than the inner planets. By comparing the mean motion of the stars observed in different parts of the sky, we are able to detect and measure the differential motion of rotation. The magnitude of the effect will depend on the average distance of the stars surveyed; because the farther our survey extends, the greater will be the difference of velocity of the outermost and innermost stars comprised in it. We can use this Oort effect, as it is called, to measure the average distance of any class of stars, provided that the stars are well distributed round the sky.

The stars which show fixed calcium lines are so remote that we cannot use any of the more elementary methods of measuring distances. But we have now two methods of finding the average distance of a class of stars which are especially appropriate to large distances, (1) by the intensity of the fixed calcium or sodium lines, and (2) by the Oort rotation effect; and we can check one against the other. Plaskett and Pearce first sorted out their stars into three groups according as the calcium lines were weak, medium or strong; these accordingly comprise the near, intermediate and distant stars. From the measured velocities they then determined the Oort effect, and thus found the average distances of the three groups. These proved to be in the order expected. This was the first check.

In calculating the Oort effect the velocities shown by the ordinary spectral lines of the stars were used—not the calcium lines which belong to the cloud. But Plaskett and Pearce also made another similar calculation using the velocities given by the fixed calcium lines. We have seen that the calcium cloud is nearly at rest relative to the mean of the stars, so that it evidently shares with them in the galactic rotation; and it should therefore show the Oort effect. Accordingly for each of their three groups Plaskett and Pearce found a distance of the cloud as well as a distance of the stars. Their measurement referred, of course, not to the whole cloud but to the part of the cloud which was performing the absorption and creating the spectral lines. If the veiling cloud between us and the star is uniform its average distance will correspond to a point half-way between us and the star. Thus the distance found for the cloud should always be half the distance of the corresponding stars. This was found to be closely fulfilled. The actual results which exhibit this are as follows:

Calcium absorption	No. of stars	Stars		Cloud	
		Rotation effect km. per sec.	Deduced distance parsecs	Rotation effect km. per sec.	Deduced distance parsecs
Low	90	10·2	600	5·0	295
Medium	79	14·5	850	6·9	405
High	43	27·5	1620	13·7	805

Let us now return to the difficulty which for a long time baffled astronomers, namely that only certain types of stars show this effect. It is really due to a chapter of accidents. Naturally we shall only detect the absorption if there is a large thickness of cloud between us and the star; so that only stars distant more than, say, 300 parsecs are eligible. Since its apparent brightness must be sufficient to allow us to examine the spectrum, and since it must be distant more than 300 parsecs (1000 light-years), the star must have very high intrinsic luminosity. That greatly restricts the possible types of stars. Then further the star must be of such a type that it does not produce the calcium and sodium lines on its own account; for in a stellar atmosphere these lines (if they occur) are strong and broad, and they may completely mask the fine sharp lines which the cosmic cloud superimposes. When both these factors are taken into consideration the limitation to the particular types is fully explained.

That this explanation is right has been proved recently by several instances in which, owing to exceptional circumstances, it has been possible to discover the fixed lines in stars of the "wrong" type. If the trouble is that the fixed line is being masked by the star's own calcium or sodium lines, it occasionally happens that we can surmount it. In some double stars the two components have extremely rapid motion; then at a certain phase the calcium lines of one component will be displaced by the motion well away to the right and the lines of the other component to the left, leaving a gap where the fixed or interstellar calcium line can show itself. This has duly been observed to happen.

III

Why calcium and sodium? I do not for a moment suppose that the cloud is composed wholly or even mainly of these two elements. But running through the list of the elements, we soon satisfy ourselves that calcium and sodium are the only reasonably abundant elements that, under the conditions of stimulation prevailing in interstellar space, could yield spectral lines observable by us. It is no accident that the cosmic cloud is betrayed by three particular spectral lines, H, K and D; these and no others are the lines which rarefied matter composed like an average sample of terrestrial matter would display.

Although we can learn a great deal about the chemistry of the heavenly bodies we have not all the advantages that a laboratory analyst has. He, if he wants to find out whether a particular element is present in a sample of material, takes care to provide the conditions of heat or electrical stimulation which are most favourable for his investigation. We have to take the conditions as we find them; and if they are not favourable for developing a particular spectrum we miss the corresponding element in our search. The worst handicap of the astronomer is that all celestial spectra are cut off abruptly at about wave-length 3000 Å., a point at which the laboratory physicist would say that spectra are just beginning to be most informative. We are in the position of a listener trying to follow a piece of music with a loud speaker that can reproduce only the bass notes. A layer of ozone high up in our atmosphere is opaque to radiation beyond the limit I have mentioned; so we lose all the treble notes in the song of the celestial atoms. Calcium and sodium have deep chesty voices and can make themselves heard.

Let us turn now to considerations of a more theoretical kind. We want to gain some idea of the density of the cosmic cloud. Various lines of argument prove that it must be extremely tenuous. One proof rests on Einstein's theory. For any given density there is an upper limit to the greatest possible extension of the cloud. For example, a globe of matter of the density of water cannot possibly be more than 400 million miles in diameter. Perhaps I had better explain why. But the worst of explanations is that they often provoke more questions than they answer, and I shall not be surprised if you find my explanation more incredible than the statement itself. It is nevertheless a sober scientific calculation due originally to Schwarzschild. By Einstein's law of gravitation a lump of matter causes a curvature of the space which it occupies. If you enlarge it you add more space of the same curvature. You can go on enlarging it until the space has curved right round and closed up; then you must perforce stop. That is what happens to the globe of water; when it has been enlarged to a diameter of nearly 400 million miles, space closes tightly up all round and there is nowhere to put any more water or anything else. Unless you have taken the precaution of immersing yourself in the water, you will be —nowhere.

Another way of reaching the same upper limit is to consider that, if the globe is large enough, its gravitation will be so intense that neither light nor anything else can escape from it; so that it will form an entirely self-contained universe. In order to support myself with authority I will give a quotation—

A luminous star of the same density as the earth, and whose diameter should be 250 times larger than that of the sun, would not, in consequence of its attraction, allow any of its rays to arrive at us; it is therefore possible that the largest luminous bodies in the universe may, through this cause, be invisible.

Perhaps you will look on this as one more illustration of the disastrous effects of the Einsteinian revolution on respectable scientific investigation, and lament the old days when the teaching of Newton, Laplace and other giants of the past kept science in the true path of sanity. But do not be in too much of a hurry to blame Einstein. The passage quoted is from Laplace's *Système du Monde* (1796). Even Newton thought that light might be subject to gravitation, and by Laplace's time it had come to be generally assumed that it was. Passing over a century during which (owing to the undulatory theory) it was generally supposed that light was not subject to gravitation, the first observational proof of the action of gravitation on light was in 1919.

The lower the density, the larger the globe that can be built. Evidently we must take the density of the cosmic gas low enough to build a cloud which can contain our whole galactic system. This condition requires that the density shall be less than 10^{-18}, that is to say one million million millionth of the density of water.

A still more stringent limit is found by considering the observed velocities of the stars. The more gravitating matter there is in the stellar system, the greater are the forces to which the stars are subjected, and the greater will be the average speed of stellar motion. By this criterion it was found that the density of the cloud could scarcely be greater than 10^{-23}; but the calculation may not be very trustworthy, since the ideas on which it was based have been somewhat modified by the discovery of the rotation of our galaxy.

These are upper limits. A more definite estimate of the average density of the cosmic cloud is obtained by considering the way in which the density of a nebula tails off into the normal uncondensed cloud. We shall see later that both the cloud and the nebulae are at a rather high temperature of the order 10,000° to 20,000°. We can make what is probably a near enough guess at the average weight of the particles; it will be considerably greater than the corresponding quantity for the interior of a star (p. 146) because the atoms are less highly ionised. Then for a nebula of given temperature and average molecular weight it is possible to calculate the way the density falls off from the centre outwards; and fortunately for us the density at large distances from the centre turns out to be nearly independent of the central density (which we should have been quite unable to estimate).

There is no definite boundary to a nebula. The density continually falls off at greater and greater distances until we come to the outskirts of the next adjacent nebula. Thus if we want to know the average density of the gas in a normal region of space, we have to ask ourselves how far on the average will it be from the centre of the nearest nebula; we may then calculate the density as though it were part of that nebula. For the calculation we require, besides the distance, only the temperature and the average molecular weight, as explained above. From the observed distribution of the nebulae it is estimated that normally the nearest nebula is 100 to 200 parsecs distant. This gives a density of 10^{-24}, in round numbers, for an average region of the cosmic cloud.

If this is the right order of magnitude of the density, the amount of matter in the cloud is roughly the same as the amount condensed into stars. This is in agreement with a theoretical study of the conditions of formation of condensations in a uniform primordial nebula, which indicates that $\frac{2}{3}$ of the matter will form condensations and $\frac{1}{3}$ will be left uncondensed.

At the centre of a typical nebula, e.g. the Great Nebula in Orion, the density must be about 10,000 times greater, viz. 10^{-20}. This is one-millionth of the density in the highest vacuum that we can create in the laboratory. So throughout the present chapter I am talking about that which—by terrestrial standards—is less than nothing.

The density will be more vivid to us if we express it in terms of atoms. A density of 10^{-24} means that there is about one atom to the cubic centimetre, if—as in the stars—the majority of the atoms are hydrogen. I suppose it is rather startling to realise that in the remote solitude of interstellar space an atom still has neighbours within an inch of it. I wanted to impress on you the extreme tenuity of the cosmic cloud; but my last statement is likely to reverse the impression, giving you a picture of the atoms swarming as thickly as a plague of gnats. The picture is true enough; but we have to remember that an atom is a most insignificant quantity of matter. A moderate smoker will in the course of a day pollute the air with a prodigious number of atoms—so many that, if we suppose them to diffuse evenly through the atmosphere all over the earth, no one will be able to draw a breath anywhere without inhaling a dozen atoms that have come from the offending pipe.

Take a cupful of liquid, label all the atoms in it so that you will recognise them again, and cast it into the sea; and let the atoms be diffused throughout all the oceans of the earth. Then draw out a cupful of sea-water anywhere; it will be found to contain some dozens of the labelled atoms. We can read a literal meaning into Macbeth's words:

> Will all great Neptune's ocean wash this blood
> Clean from my hand? No, this my hand will rather
> The multitudinous seas incarnadine.

One atom per cubic centimetre does not amount to much. A portion of the cosmic cloud as large as the earth could, if compressed, be packed in a suitcase and easily carried with one hand.

IV

The most paradoxical thing about the cosmic cloud is that it is intensely hot. We often speak of the intense cold of interstellar space. It is quite true that far away from the sun, at an average point in our galaxy, the temperature of any solid or liquid body would fall to $-270°$ C., or $3°$ above absolute zero. That is the temperature that would be indicated by a thermometer; it is the degree of cold which the human body would feel, if feeling could be imagined under such conditions. But the diffuse cloud, by reason of its diffuseness, contrives to keep warm in the same conditions.

Crossing any region of space there is a certain amount of heat radiated by the stars. Altogether it amounts to about the heat of a candle 100 yards away. You can imagine that it would be a bit chilly to sit out in space trying to warm yourself by a candle 100 yards away. If the human body could store up all the heat received minute by minute from the candle, you would in the end become warm; but matter is so constituted that it dissipates any heat contained in it, and as soon as the temperature has risen to $3°$ absolute this loss becomes sufficient to neutralise the gain.

The reason why the diffuse cosmic gas reaches a higher temperature is that it has less opportunity of losing the heat it collects. The heat of a gas is the energy of motion of its particles (molecules, atoms or free electrons), and the time when there is a risk of losing some of this energy is during a collision of two particles. In air under ordinary conditions each particle undergoes some thousands of millions of collisions every second. In the cosmic cloud an atom encounters another atom about once a year; it has, however, as a milder excitement a collision with an electron about every five days. Owing to this rarity of collisions, the process of loss of heat which operates in ordinary solid bodies is rendered practically idle in the cosmic cloud.

That, however, is not the whole secret of the high temperature of the cloud. The processes by which a body loses heat are closely bound up with the processes by which it acquires heat, so that the argument cuts both ways. The collisions are an opportunity for gathering in the radiant heat that is passing, as well as for losing it; and owing to their rarity the gas lets most of the radiation pass through without being warmed by it; that is to say, it is highly transparent. So if we imagine a piece of the cosmic cloud and a solid meteorite each sitting in front of a candle 100 yards away and trying to get warm, it is not immediately obvious which will have the advantage. The cosmic cloud secures very little heat but it does not easily lose what it does secure; the meteorite secures all that comes its way but parts with it easily. All we can say is that the mechanism, which determines what temperature the meteorite will take up, is practically out of action in the cosmic gas; so that there is no reason for them to have the same temperature. In the cosmic gas the field is left clear for a secondary mechanism, unconnected with collisions, to take control of the temperature.

This second mechanism is the "photo-electric effect" (p. 37). A quantum of light (of sufficiently high frequency) falling on an atom causes an electron to shoot away at high speed. In interstellar space the star-light is continually causing this ejection of high-speed electrons. We may say that an electron gas at high temperature is being generated—high temperature because of the high speeds. The electrons are ultimately captured again so that the electron gas is disappearing as fast as it is generated; but being always generated at high temperature it warms up the cloud.

It is not possible here to go at all deeply into the theory; but the important point is that by the laws of quantum theory the speed of ejection of the electrons, and therefore the initial temperature of the electron gas, depends on the *quality* and not on the *quantity* of the stellar radiation. It is therefore the same in the depths of interstellar space as in the close neighbourhood of the stars; and the temperature is in fact not far short of the surface-temperature of the hottest stars responsible for the radiation. Not even the quantum theory provides something for nothing, and quantity must tell in another way. The very low intensity of star-light in interstellar space does not reduce the temperature of the electron gas, but it makes its generation a very slow business. The slowness, however, does not matter in this connection, since all other ways of heating or cooling the cloud have practically stopped and there is no competitor to outstrip. So far as we can estimate the cosmic cloud will take up a temperature of the order 15,000°.

Let us now summarise the results so far reached. We started with the direct observational proof that there is an interstellar gas which gives the H and K lines of calcium and the D line of sodium in the spectra of distant stars, these lines not being attributable to the stars themselves on account of the difference of motion that is indicated. We then attacked the problem in a different way, and by an independent theoretical argument concluded that there should exist interstellar matter of density about 10^{-24} and temperature about 15,000°. It remains to connect the two investigations, and examine whether the sodium and calcium contained in a cloud of this density and temperature would give absorption lines of the intensity which we actually observe. This requires that we should examine the state of the atoms; for atoms give different spectral lines according to their state of ionisation. For example, calcium is a divalent element with two rather loosely attached electrons. Under the conditions above stated we find that the great majority of the calcium atoms will be without these two electrons; they have gone off to form part of the electron gas to which I have referred.

Now the calcium atom with two electrons missing gives

no observable spectrum; it is not these atoms that we are concerned with. The H and K lines are produced by calcium atoms with one electron missing. About 1 in 800 of the calcium atoms will be in this state. Complete calcium atoms are very rare in the cloud—about 1 in 50,000,000; the rarity explains why we do not observe in the spectrum of the cloud the lines of un-ionised calcium.

Calcium is a fairly abundant element forming about 1 per cent. of the whole mass of the earth. If we allow the same proportion in the cloud, and remember that only $\frac{1}{800}$ of the atoms are in a state to cause absorption of H and K lines, we find that there is about one active calcium atom in a cubic yard of cosmic cloud. Consider now a star 1000 light-years away. We see it across a screen 1000 light-years thick containing one absorbing atom per cubic yard. What intensity of absorption line will such a screen produce? The physicist is able to answer this question from his experimental and theoretical knowledge; and when we compare his calculation with the intensity (width and blackness) of the fixed H and K lines that we actually observe in a star 1000 light-years away, the agreement is as close as could be desired.

Unfortunately this agreement for calcium is marred by a complete disagreement for sodium which we are unable to explain. The D line is produced by the *complete* sodium atom; but in the conditions that we have calculated for the cloud the sodium ought to be nearly all ionised, and complete atoms should be far too rare to give the absorption lines that we observe. Even if the cloud consisted entirely of sodium there would still not be enough complete sodium atoms. And so I have to leave my story without a happy ending. But perhaps after all it *is* a happy ending that stimulates us to pursue farther our investigations because there is still something fundamental to be found out.

V

Although the cosmic cloud is generally invisible there are denser patches which are faintly luminous. These are the gaseous nebulae of which an example is shown in Plate 2. The most interesting part of the study of gaseous nebulae deals with the origin and nature of their light.

We are familiar with bodies such as the sun which shine by their own light, and with bodies such as the moon which shine by borrowed light. A gaseous nebula is in a sense intermediate. The nebula is dependent for its light on the stars which lie in the midst of it; but it does not simply reflect their light; their radiation falls on the atoms of the nebula and stimulates them so that they emit light of a different kind. To use the recognised term for this process, the nebula is *fluorescent*. It is only the stars of very high temperature that can cause a nebula to shine; the sun would not be capable. So even the densest portions of the cosmic cloud will remain dark unless there are high-temperature stars in the neighbourhood. We may suspect that the dark obscuring nebulae (Plate 3) are similar to the luminous nebulae but lack the stimulating stars. It is, however, very difficult to account for their opacity if they consist of gas alone; and for that reason astronomers nowadays usually look on them as clouds of dust or meteoric matter. Whatever be the solution, there is an intimate association between the obscuring nebulae and the luminous nebulae; for we often see in the same nebula luminous portions which grade continuously into dark obscuring portions.

Keeping to the luminous nebulae, their spectrum is, as we should expect, that of a practically transparent layer of gas; that is to say, it is a spectrum consisting of bright lines. The spectrum of hydrogen is a prominent, but by no means the most prominent, feature of the spectrum. Ionised helium, i.e. atoms of helium which have lost one satellite electron,

PLATE 3

Mount Wilson Observatory

DARK NEBULOSITY

The Horse's Head in Orion

can also be recognised. The rest of the spectrum consists of lines entirely unknown in the laboratory. Most of the visible light comes from two green lines whose source has been named *nebulium*. The photographic light contains another very prominent line whose source has not been specially named. Named or not, the light of the nebulae is for the most part like nothing on earth.

The modern theory of the sequence of the atomic numbers of the elements (p. 30) leaves no room for new elements until we reach very high atomic numbers. Up to the point where the first gap occurs, the physicist would be almost as surprised to discover a new element as the mathematician to discover a new integer. Thus for many years astronomers have been convinced that nebulium is an *alias* of some very familiar element, and that the sources of the other unknown lines are likewise familiar substances. The problem was how to force some familiar element to emit the strange light which it is so reluctant to give in the laboratory. Laboratory treatment of atoms is still somewhat crude. Our method of making an atom work is to knock it about; and if it does not do what we want, knock it still harder.

But is it likely that this treatment will bring out light of the kind emitted in the nebulae? In the nebulae the atoms have a very quiet life. We have seen that in the cosmic cloud the only break of monotony is an encounter with an electron about once a week. In the thousand-fold denser nebula things are speeded up proportionately; but even so, to an atom whose natural periodicity is of the order 10^{-9} seconds, the interval between encounters must seem almost an eternity.

Let us look at it another way. We can measure roughly the amount of light emitted by a luminous nebula, e.g. the Orion Nebula, and we can express the result as a certain number of quanta (or photons) emitted per second. We know also the size of the nebula, and the theory described in this chapter has given a general idea of the density. We can therefore estimate the total number of atoms in the nebula. Hence we can calculate how often on the average each atom is called upon to emit a photon. We find that its turn comes round about once a century.

The secret of nebulium was discovered by I. S. Bowen in 1927. He found that the strange light was due to what are known theoretically as "forbidden transitions". We have seen that there are a number of possible orbits for a satellite electron, and that light is emitted when the electron jumps from an orbit of higher energy to an orbit of lower energy. But the electron does not jump indiscriminately. It is as though the orbits were connected by cross-passages; some pairs of orbits have a cross-passage, others have not. It may happen, for example, that an electron in orbit No. 3 can drop to No. 2 or to a still lower orbit No. 1, but it will not drop from No. 2 to No. 1. In that case the passage from No. 2 to No. 1 is called a forbidden transition. The theory of the atom has furnished us with rules that determine which transitions are forbidden.

It was realised that the transitions are only *relatively* forbidden. The electron in the above example *can* drop from orbit No. 2 to No. 1, only the chance of its doing so in any reasonable time is small; and if it does not act quickly, it will be whisked out of orbit No. 2 by the collisions and absorptions that are continually occurring in terrestrial conditions. Bowen realised that in a nebula an electron, which had been knocked up into orbit No. 2 with only the orbit No. 1 below it, would ultimately have to make the forbidden transition. There being nothing to disturb or release it, it would remain a prisoner in orbit No. 2 until its obstinacy gave out. The unfamiliar lines in the nebular spectrum correspond to forbidden transitions, and for that reason they are only emitted in extremely quiescent conditions such as prevail in a nebula.

The proof lies in the identification of the lines. Nebulium is doubly ionised oxygen. All the other conspicuous lines in the nebulae whose origin was previously unknown are forbidden lines either of singly or doubly ionised oxygen or of singly ionised nitrogen.

How do we know what are the forbidden lines of oxygen if we cannot ourselves persuade oxygen to produce them? Except in one specially simple case we cannot calculate the spectrum of an atom by pure theory, and we are reduced to measuring wave-lengths experimentally. But there is no need to measure the wave-lengths of all the lines in the spectrum; when we have measured a certain number, we can calculate the rest. The rule of calculation, which is a simple one, is well known in quantum theory. If the calculated line is not forbidden, we can observe it and so verify the rule; but the same rule also enables us to calculate the wave-lengths of the forbidden lines which we cannot observe.

Thus Bowen recognised the nebulium spectrum as a spectrum of oxygen although terrestrial oxygen had never been known to produce it. He recognised it as the spectrum, known in theory, which under ordinary circumstances oxygen is forbidden to produce. It is still unproduced in the laboratory. Even if we could secure sufficient quietude for the atoms, we could scarcely expect to detect the light. For the atoms, as we have seen, take their time over emitting it and cannot be made to work faster than, say, once a second. That means that the light is generated very feebly, and a source of astronomical dimensions is required to yield an appreciable quantity.

So the source of

> The light that never was on sea or land

is a familiar enough substance. It is oxygen and nitrogen—or, if you like, common air.

CHAPTER X

THE EXPANDING UNIVERSE

> Or if they list to try
> Conjecture, he his fabric of the heavens
> Hath left to their disputes.   MILTON, *Paradise Lost.*

I

THIS chapter describes a material system on the largest scale yet imagined, namely the system of the galaxies. Let us first understand what a galaxy is. The following is a recipe for making galaxies: Take about ten thousand million stars. Spread them so that on the average light takes three or four years to pass from one to the next. Add about the same amount of matter in the form of diffuse gas between the stars. Roll it all out flat. Set it spinning in its own plane. Then you will obtain an object which, viewed from a sufficient distance, will probably look more or less like the spiral nebula shown in Plate 4.

The evidence is now considered conclusive that the spiral nebulae (not to be confused with the *gaseous nebulae* considered in the preceding chapter) are immense systems of stars. They are presumably the units that we have to deal with in a survey of the universe as a whole. We, of course, live in one of these galaxies, of which the sun and all ordinarily recognised stars are members. We call it *the Galaxy*. Being inside it we do not get so good a general view of it, and it is difficult to compare it with the other galaxies. From an observational point of view there is some doubt whether our galaxy is a normal specimen; it appears to be outstandingly large. But the tendency of recent investigations has been to level up things, and make our galaxy seem less abnormal.

PLATE 4

Mount Wilson Observatory

SPIRAL NEBULA

M. 51 in Canes Venatici

Judging by sample counts in different parts of the sky there are some millions of these galaxies visible with our largest telescopes; and goodness knows how many there are beyond their range. Or rather—I think I also know more or less. But that is all theory, and you will have enough of that later on.

The spiral nebulae or galaxies will now be our units. Just as the chemist generally takes an atom as his unit and does not need to trouble about anything smaller, dealing as he does with aggregations of myriads of atoms, so we shall take the galaxies as our atoms, not recognising anything smaller, and discuss an aggregation of, say, a billion galaxies which, it seems, constitutes the universe.

Let us now see where we have got to in the scale of size:

	Miles
Distance of sun	93,000,000
Limit of solar system (Orbit of Pluto)	3,600,000,000
Distance of nearest star	25,000,000,000,000
Distance of nearest galaxy	6,000,000,000,000,000,000
Original circumference of the universe	40,000,000,000,000,000,000,000

The last entry is here rather premature, but we shall refer to it later (p. 218).

I have sometimes heard complaints that astronomical numbers are too large to comprehend. I must confess that the numbers in the present chapter are rather out of the ordinary. But I cannot see why anyone should find a difficulty with the numbers occurring in the lesser astronomical systems. They are just the sort of figures with which our economists and politicians are always dealing, quoted any day in the newspapers. People say that they cannot realise these big numbers. But that is the last thing anyone wants to do with big numbers—to realise them. Do you suppose that, as Budget Day approaches, the Chancellor of the Exchequer throws himself into a state of trance in which he can visualise and gloat over 800,000,000 sovereigns, or notes, or commodity values, or whatever they are, that he is about to amass? His only concern is to make quite sure that, although neither 800,000,000 nor 8,000,000,000 is a number that can possibly be "realised", he does not forget which is which. The purpose of the foregoing table of distances is that we may keep the different scales distinct in our minds; in particular we have to note that it is a very big step up in scale when we pass from a system of stars to a system of galaxies.

How do we find the distances of the galaxies? For a start, we can actually photograph some of the brightest of the stars in the nearest galaxies. Their distance makes them appear very faint; and the faintness is a measure of the distance, provided that it is known how bright these same stars would appear at a standard distance, and provided also that there is no intervening fog. If all motor cars were equipped with lamps of standard power, it would be possible to tell how far off each car was by carefully measuring the apparent brightness of its lamps. The galaxies carry many lamps, and among these we can recognise some that are known to be of standard power, namely the Cepheid Variables. If we observe in one of the galaxies a star varying in the Cepheid manner with a period of 10 days, we look up in our list the standard light-power of a Cepheid of 10 days' period (as determined from the measurements of such stars in our own neighbourhood), and the distance of the galaxy is then immediately deduced. E. P. Hubble has measured the distances of a few of the nearest galaxies in this way. For greater distances less satisfactory methods are employed, and I daresay the distances assigned to the remoter galaxies are a bit doubtful. However, we think that they give a reasonably good idea of the system.

We can also determine how fast a galaxy is moving towards or away from us in the line of sight by measuring the Doppler shift of the lines in its spectrum—a method applicable to all luminous objects whose spectra are not entirely featureless. For the stars of our own system it is rare to find velocities above 100 miles per second; but the velocities of the external galaxies are generally much larger, amounting to hundreds or thousands of miles per second.

There is a remarkable feature about these motions. The more distant the galaxy, the faster its motion. Moreover, the galaxies are almost unanimously running away from us.

The nearest of the external galaxies is about a million light-years away. At present the farthest limit of our survey is about 150 million light-years. There is a spiral nebula in the constellation Gemini whose distance is such that the light waves had to start 150 million years ago in order to reach us to-day; this nebula is running away from us at 15,000 miles a second. Intermediate galaxies recede at less speed, the speed being, as nearly as we can tell, proportional to the distance. The system undoubtedly extends farther than 150 million light-years; but the more remote galaxies are so faint that it has not yet been practicable to make the measurements necessary to determine their speeds. [Announcement has just been made (November, 1934) of a still more remote nebula in Boötes receding at 24,300 miles a second.]

There are five exceptions to the rule that the external galaxies are receding from us; but the exceptional behaviour is confined to galaxies in our immediate neighbourhood and is probably not of real importance. By the law of proportionality of speed to distance the closest galaxies, distant from one to two million light-years, should have an outward speed from 100 to 200 miles per second. This is scarcely large enough to predominate over various accidental causes which may be operating; so that no great attention need be paid to an occasional reversal in this region. As soon as the distance effect becomes too large to be masked, the recession of the galaxies is unanimous.

II

If all the galaxies are going away and none are moving inwards, there will come a time when the region that we now survey is vacated. I cannot say that that will make much difference to earthly affairs; but the astronomers of that future time will lose one of the most fascinating and beautiful features of the heavens. Nor is that date so very far off. I do not want to be unduly alarmist; but the nebulae double their distances from us every 1300 million years, and astronomers will have to double the apertures of their largest telescopes every 1300 million years merely to keep up with their recession. Seriously, 1300 million years cannot be looked upon as a long period of cosmic history; it is about the age assigned to the oldest terrestrial rocks; and it is certainly a novel idea that anything much can have happened to the vast system of the universe within geological times. It implies that the time-scale of change and evolution is much shorter than we were inclined to think a few years ago. We have to speed up the evolution of the stars in order to harmonise with it.

The running away of the galaxies does not mean that they have a kind of aversion from us. They are not avoiding our own galaxy; it is not important enough for that. If we consider carefully the observational law, that the speed of recession is proportional to the distance, we see that the galaxies are separating away from one another just as much as they are separating away from our galaxy. An even dilatation of the whole system is occurring. If this lecture-room were to expand to twice its present size, the seats all separating from one another in proportion, you would at first think that everyone was moving away from you. But everyone else would be having the same experience. It is that kind of expansion which is occurring in the system of the galaxies. Since the rate of mutual recession, or rate of increase of distance from any galaxy to any other, is proportional to the distance, all distances take the same time to become doubled, viz. 1300 million years.

So the system of the galaxies is expanding as a gas expands, its atoms getting farther and farther apart from one another. (You will remember that in this super-physics of the universe the galaxies are our indivisible units or atoms.) If the astronomers are right, it is a straightforward conclusion from the observational measurements that the system of the galaxies is expanding—or, since the system of the galaxies is all the universe we know—that the universe is expanding. There is no subtlety or metaphysics about it. Except that the system concerned is of unaccustomedly large dimensions, it is easy enough to apprehend.

But are we sure of our observational facts? Scientific men are rather fond of saying pontifically that one ought to be quite sure of one's observational facts before embarking on theory. Fortunately those who give this advice do not practise what they preach. Observation and theory get on best when they are mixed together, both helping one another in the pursuit of truth. It is a good rule not to put overmuch confidence in a theory until it has been confirmed by observation. I hope I shall not shock the experimental physicists too much if I add that it is also a good rule not to put overmuch confidence in the observational results that are put forward *until they have been confirmed by theory*.

So in starting to theorise about the expanding universe I am not taking it for granted that the observational evidence which we have been considering is entirely certain. That is what I want to find out—whether theory confirms it. At the start we have a certain reluctance to accept these observational results at their face value. It is not the expansion of the universe but the *rapid* expansion which makes us look at these observational results very critically; for if they are true they play havoc with our former ideas of the time-scale of evolutionary development. If the speeds found for the spiral nebulae are genuine, there is no escape from this rapid expansion.

But are they genuine? It is scarcely true to say that we *observe* these velocities of recession. We observe a shift of the spectrum to the red; and although such a shift is usually due to recession of the object, it is not inconceivable that it should sometimes arise from another cause. I can only say that nothing in our knowledge of physics as it stands to-day gives any hint of an alternative cause for the red-shift of the nebular spectra; there would have to be some profound modification either in the theory of light or in astronomical conclusions generally. It can be objected that our knowledge is incomplete and that there is a possibility of unforeseen developments; but that might be urged against most of our scientific conclusions, and it is misleading to remember it suddenly in one particular connection. In a recent *General Catalogue of Radial Velocities* the determinations for all other objects are given in a column headed "Radial Velocity", but for the spiral nebulae the column is headed "Apparent Radial Velocity". To many that will seem commendable caution; but to me it seems like the caution of the minister who wrote to his wife "I shall be home (D.V.) on Friday; and in any case by Saturday".

In introducing theory, I must emphasise that it was theory that first suggested a systematic motion of recession of the spiral nebulae and so led to a search for this effect. The theoretical possibility was first discovered by W. de Sitter in 1917. Only three radial velocities of nebulae were known at that time, and they somewhat lamely supported his theory by a majority of 2 to 1. Since then it has been possible to investigate the more remote nebulae whose support is so far unanimous; this progress has been mainly due to V. M. Slipher at the Lowell Observatory and to M. L. Humason at Mount Wilson Observatory. The linear law of proportionality between speed and distance was found by E. H. Hubble. Meanwhile the theory has also developed, and it has taken the form especially associated with the names of A. Friedman and G. Lemaître.

The theory of relativity predicts the existence of a certain force which we call *cosmical repulsion*. It is directly proportional to the distance of the object concerned. It is so weak that we can leave it out of account in discussing the motions of the planets round the sun or indeed any motion within the limits of our own galaxy. But since it increases proportionately to the distance we shall, if we go far enough, find it significant. Will it have become significant at the distance of the spiral nebulae? The theory of relativity could not say; it did not predict the magnitude of the force. It could only suggest that a search be made as to the motions of these remote objects; and if a general running away, such as would be produced by a repulsive force, were discovered, it might well be the manifestation of this cosmical repulsion.

In the foregoing paragraph I have said that the repulsion is proportional to the distance of the object. Distance from what? *From anywhere you like.* We take it to be distance from the earth, or rather from our galaxy—since the galaxies are our indivisible atoms. But an observer in another galaxy can take it to be distance from him. It does not matter; we shall all obtain the same results so far as anything observable is concerned. Cosmical repulsion is a dispersing force tending to make a system expand uniformly—not diverging from any centre in particular, but such that all internal distances increase at the same rate. That corresponds precisely to the kind of expansion we observe in the system of the galaxies.

I have said that relativity theory predicts a force of cosmical repulsion. When using its own technical language, relativity theory does not talk about anything so crude as force; it describes the phenomena by means of curvature of space-time. But for practical purposes the curvature of space-time involved in gravitational effects is very nearly equivalent to the Newtonian force of gravitation; and the force of cosmical repulsion is similarly a translation into Newtonian language of another curvature effect demanded by relativity theory. These translations must, of course, be used with caution and not pressed to apply in extreme circumstances. There would be no object in the recondite phraseology of the theory if a familiar translation were equally satisfactory for all purposes. However, the actual relativity effect is represented with sufficient accuracy by a force of cosmical repulsion at any rate up to the greatest distances that we actually observe.

Cosmical repulsion is not the only force at work. The galaxies exert on one another their ordinary gravitational attraction approximately according to Newton's law. This makes them tend to cling together. So we really have a contest of two forces, Newtonian attraction trying to keep the universe together and cosmical repulsion trying to scatter it. If our theory is right cosmical repulsion must have got the upper hand, because the galaxies are actually being scattered. Having got the advantage, cosmical repulsion will keep it; because, as the nebulae become farther apart, their mutual attraction will become weaker and offer less opposition to the scattering force.

In connection with cosmical repulsion we define an important constant of nature, called the *cosmical constant*, i.e. the amount of the cosmical repulsion at unit distance from the observer. This constant is generally denoted by λ. Ordinary relativity theory does not foretell the magnitude of λ, or even its sign (plus or minus). All that it insists is that λ is not zero; for the theory would then cease to be a relativity theory.*
It was a defect of Einstein's original theory, first remedied

* The cosmical constant expresses a relation of scale between two types of phenomena. So long as it is expressed by any number, however small, the relation remains recognised. But if it is expressed by zero the relation is broken.

by H. Weyl, that it implied the existence of an absolute standard of length—a conception as foreign to the relativistic point of view as absolute motion, absolute simultaneity, absolute rotation, etc. To set $\lambda=0$ implies a reversion to the imperfectly relativistic theory—a step which is no more to be thought of than a return to the Newtonian theory.

Accordingly the adopted value of the cosmical constant is generally determined from the observed rate of recession of the galaxies. Such a determination is necessarily provisional since it assumes that the large receding velocities are due to cosmical repulsion and not to other causes. We shall see later, however, that the theory of the cosmical constant can be approached in another way which gives a definite determination of its value independent of the astronomical evidence, and thereby provides a check on the whole theory.

III

The reader who has followed articles and discussions on this subject may have wished to interrupt me with a question. "You have been describing the expansion of a big system of galaxies which forms the material universe; but is not the 'expansion of the universe' understood to mean something more than this—not just the pushing farther back of the boundaries of a material system, but an expansion, an inflation, of space itself?" That is true, and I must say a little about the expansion of space although the idea is more difficult to follow.

A new phenomenon is naturally considered and described in relation to the general physical theories prevailing at the time. Thus a new atomic phenomenon would nowadays be described according to wave mechanics, although it might happen to have little or no concern with the *distinctive features* of wave mechanics, and a classical description and explanation would be adequate. Although it is useful to recognise that the phenomenon does not take us beyond the limits of classical theory, we shall not get full value out of it as a contribution to the general development of science unless we weave it into the most up-to-date point of view. The physicists of a hundred years ago might well be surprised to learn that the positron is considered to require an intricate explanation barely comprehensible to anyone but a mathematician, and that it is sometimes even claimed to be a confirmation of modern ideas; they would see in it rather a confirmation of their own commonsense view that there are two electric fluids with perfectly symmetrical properties capable of cancelling one another. In the same way we have here encountered a new astronomical phenomenon which, on the face of it, is nothing more than an ordinary expansion of a material system. Modern theory is only involved to the extent of suggesting the cause of scattering (cosmical repulsion), which otherwise has to be postulated *ad hoc*. The reason why we do not rest content with this description is that the system is on a very much larger scale than any system whose expansion has previously been studied; and the large scale brings out certain differences between the modern scientific outlook and the classical outlook which would be insignificant in a smaller system. It is in this connection that the idea of an expansion of space occurs. In a briefer account of the expanding universe it would, I think, be justifiable to omit all reference to expanding space; just as in briefly introducing the positron I have not referred to Dirac's theory of it as an occasional vacancy in an infinitude of occupied negative energy-levels.

We have seen that the speed of recession of a galaxy is proportional to its distance. At 150 million light-years the speed is 15,000 miles a second; at 1500 million light-years the speed should be 150,000 miles a second. But we cannot go on indefinitely like that. At 1900 million light-years we get 190,000 miles a second, which is greater than the speed of light; so that we are obviously heading for trouble. The trouble indeed is so near that if Dr Hubble had been armed with a 1000-inch telescope instead of a 100-inch he would probably have landed us in it already.

Einstein about 1916 seems to have had a premonition that if we include very great distances in our scheme of things we are asking for trouble. That queer quantity "infinity" is the very mischief, and no rational physicist should have anything to do with it. Perhaps that is why mathematicians represent it by a sign like a love-knot. Einstein therefore adopted a type of space in which there are no distances beyond a certain amount, just as on the earth's surface there are no distances greater than 12,000 miles. We have just seen that there is trouble in store for us if we go out to too great a distance in the system of the galaxies; but Einstein has taken the precaution of closing up the universe so that we cannot wander too far.

It is not an accident that the closure of space saves the situation. We have seen that the force of cosmical repulsion is, like the force of gravity, an approximate equivalent in familiar language of the curvature of space-time in relativity theory. The closing up of space, so that its volume is finite and distances cannot exceed a finite limit, also results from the curvature. Thus the extent of space and the magnitude of the force of cosmical repulsion are proportioned to one another, both depending on the same cosmical constant; and their relation is such that the anticipated trouble cannot arise.

To sum up: if we accept the force of cosmical repulsion offered by relativity theory, we should for consistency accept the finitude of space that goes along with it on that theory. It is quite true that the latter scarcely affects the problem with which we are primarily engaged, viz. the expansion of the system of the galaxies, unless we contemplate distances ten times greater than those yet observed; and we have as yet no evidence that the system extends so far as that. But when we go on to consider the structure and evolution of the universe as a whole, we naturally appeal to the complete self-consistent theory.

If the system of the galaxies extends throughout closed finite space, it can only expand if the space itself expands. That is how we are led to contemplate expanding space as well as an expanding material system.

IV

To simplify things we shall suppose that the distribution of the galaxies is uniform throughout space. Space will then be spherical; that is to say, it will be like the surface of a sphere, only with one more dimension which you must imagine as best you can. More technically it is like the three-dimensional surface or boundary of a hypersphere in four dimensions. I say it is *like* the surface of a hypersphere—that we can make the simplest kind of map of space by drawing it on a hypersphere. I do not say it *is* the surface of a hypersphere, for the hypersphere is only the scaffolding of the map. Since the idea of the map is that the whole external world corresponds to the surface of the hypersphere, the interior and exterior of the hypersphere can have no objective counterpart.

A being limited to the surface of a sphere will, if he goes straight ahead turning neither to the right nor to the left, ultimately find himself back at his starting point. Similarly you, limited to a three-dimensional space which is like the surface of a hypersphere, will, if you go straight ahead, arrive back at your starting point. I cannot say exactly how far you will have to go, but the distance is not less than 6000 million light-years; it may be five or ten times as much, but I think not more. Only you had better hurry up, because the universe is expanding, and the longer you put it off the farther you will have to go. As a matter of fact it is too late to start now even if you travel with the speed of light. Adopting my minimum figure of 6000 million light-years, it will take you 1500 million years to go a quarter way round. But we have seen that the expansion is such that distances are doubled in 1300 million years. So that the remaining three-quarters of your circuit, instead of being 4500 million light-years, will now have become more than 9000 million light-years. You are farther off than when you started. One is reminded of the effort of Alice and the Red Queen—

"Well, in *our* country," said Alice, still panting a little, "you'd generally get to somewhere else—if you ran very fast for a long time, as we've been doing."

"A slow sort of country!" said the Queen. "Now *here*, you see, it takes all the running *you* can do, to keep in the same place. If you want to get somewhere else, you must run at least twice as fast as that."

Spherical space presents many such curiosities, but we shall not here linger over them.* For the most part they do not lead to anything that could come under practical observation.

In contemplating the dispersing system of the galaxies we cannot refrain from asking, What has it come from? Where is it going to? To the latter question there is, so far as we can see, only one answer. The system will go on dispersing for ever—the galaxies scattering more and more widely. Cosmical repulsion increases the distances between the galaxies, but it does not make an individual galaxy grow any larger. This is because the dispersal of a system only occurs if the repulsion exceeds the countervailing gravitational attraction of the parts of the system. In the galaxies and other smaller systems gravitational attraction always predominates. So although we are parting company with the millions of galaxies around us, we shall keep with us a

* See *The Expanding Universe*, Ch. III.

galaxy of some 10,000 million stars, which a few years ago would have been considered a fairly commodious universe.

It is more difficult to decide what the universe started from. For my part I choose the hypothesis which provides the most quiescent and orderly beginning of things. If you prefer the view (favoured by Lemaître) that the universe started with the thunder of an explosion, there is nothing in our present knowledge to gainsay you; only it seems inartistic to give a universe, built to contain a natural cause of expansion, an additional shove off at the start.

We have seen that Newtonian attraction and cosmical repulsion are two opposing forces. It would seem that in the initial state of things these two forces just balanced, so that ideally the universe might have remained in this embryo state for untold ages. But it can be shown that the equilibrium is unstable. If cosmical repulsion once gets the upper hand it will keep it (p. 214), and the universe will go on expanding; similarly if Newtonian attraction gets the upper hand it will keep it, and the universe will go on contracting. Sooner or later some slight disturbance of perfect equilibrium was bound to occur and cause the universe to topple off its balance one way or the other. Several investigators have tried to examine whether there was some definite cause deciding that the universe should fall into a state of expansion rather than contraction; but no very decisive conclusion has been reached.

According to observation the speed of recession of the spiral nebulae is (in round numbers) 500 km. per sec. per megaparsec;* that is to say, those at 1 megaparsec distance recede at 500 km. per sec., those at 10 megaparsecs distance recede at 5000 km. per sec., and so on. From this datum we can by Lemaître's theory evaluate the cosmical constant and several important characteristics of the universe. Assuming that the universe started from the state of balance described

* 1 megaparsec = 3·26 million light-years.

above, its initial radius was about 1000 million light-years. It has since expanded; but the present radius can only be found by introducing very precarious estimates of the average density of matter in the system of the galaxies.

We also deduce by Lemaître's theory that the total amount of matter in the universe is about 10^{22} times the sun's mass. If an average galaxy contains 10,000 million stars, this would provide for about a billion (10^{12}) galaxies. Another form of the result is that there are 10^{79} protons and as many electrons in the universe. That is quite a useful thing to know. We shall find confirmation of this number in Chapter XI.

I am told that this is not the first attempt to compute the number of particles in the universe. There is an earlier calculation by Archimedes.*

There are some, king Gelon, who think that the number of the sand is infinite in multitude; and I mean by the sand not only that which exists about Syracuse and the rest of Sicily but also that which is found in every region inhabited or uninhabited. Again there are some who, without regarding it as infinite, yet think that no number has been named which is great enough to exceed its multitude.... But I will try to show you by means of geometrical proofs, which you will be able to follow, that, of the numbers named by me, and given in the work which I sent to Zeuxippus, some exceed not only the number of the mass of sand equal in magnitude to the earth, but also that of a mass equal in magnitude to the universe.

The calculation proceeds by steps, from sand-grains to poppy-seeds, to finger-breadths, to stadia, to the diameter of the earth, to the diameter of the universe according to "the common account, as you have heard from astronomers", and finally to the many times greater universe recently advocated by Aristarchus. Archimedes concludes that "a sphere of the size attributed by Aristarchus to the sphere

* Sir Thomas Heath, *The Works of Archimedes*, pp. 221–232.

of the fixed stars would contain a number of grains of sand less than 10^{63}".

I will not enter into controversy with my venerable rival. I feel that we are drawn together by his concluding remark—

I conceive that these things, king Gelon, will appear incredible to the great majority of people who have not studied mathematics.

V

The conception of the expanding universe seems to crown the edifice of physical science like a lofty pinnacle. Or perhaps its strange fantastic character suggests that it would be more aptly compared to a gargoyle. But for my part I do not look on it either as a pinnacle or a gargoyle. I believe that it is one of the *main pillars* of the edifice.

The cosmical constant is the agent behind the phenomenon of the recession of the galaxies. But it is also the agent behind a great deal more. A few years ago I became strongly convinced that in these astronomical discoveries in the remoteness of space we had picked up the key to the mysteries of the proton and electron. All that I have since been able to work out confirms my conviction. In spherical space those who start off in one direction must ultimately meet those who started off in the opposite direction; so in science astronomers who went in search of the inconceivably great are now meeting atomic physicists who went in search of the inconceivably small.

The same cosmical constant found from the motions of the galaxies can also be found from the properties of electrons and protons studied in the laboratory. We have thus two independent determinations of the cosmical constant which are found to check one another as closely as could be expected. The theory of the laboratory determination—the formula giving λ in terms of the other better known constants of Nature—will be treated seriously in Chapter XI. Here I introduce only preliminary considerations.

When we assert that the universe expands, what is our standard of constancy? There is no particular subtlety about the answer; the expansion is relative to the standards that we ordinarily employ. It is relative to the standard metre bar, for example, or to the wave-length of cadmium light which is often suggested as a more ideal standard, or to any of the linear dimensions associated with atoms, electrons, etc. which are regarded as "natural constants" in atomic physics. But if the universe is expanding relatively to these standards, all these standards are shrinking relatively to the universe. The theory of the expanding universe is also the theory of the shrinking atom. Thus we cannot detach the theory of the universe from the theory of the atom. We must not think of the cosmical constant as an agent which manifests itself only in the super-system of the galaxies and is insignificant in the atom and other small-scale systems. It manifests itself in a relation (of size) between the super-system of the galaxies and small-scale systems, and it is no more a characteristic of one end of the relation than of the other. Thus we ought to be able to approach the cosmical constant through the theory of the atom (or more explicitly through those equations of quantum theory which determine the extension of small-scale systems) as well as through the theory of the universe.

According to the principle of relativity we can only observe and have knowledge of the relations of things. So when we refer to the properties of any object we must always have a comparison object in mind. If we speak of its velocity, we mean its velocity relative to some comparison object or set of landmarks. If we speak of its size we must have some standard extension to compare it with. Imagine yourself to be quite alone in the universe so that there is nothing to compare yourself with—and then try to tell me how large you are. You cannot. You have no size unless something else exists for you to be larger or smaller than.

So in any statement of physics we always have two objects in mind, the object we are primarily interested in and the object we are comparing it with. To simplify things we generally keep as far as possible to the same comparison object. Thus when we speak of size the comparison object is generally the standard metre or yard. Since we habitually use the same standard we tend to forget about it and scarcely notice that a second object is involved. We talk about the properties of an electron when we really mean the properties of an electron and a yard-stick—properties which refer to experiences in which the yard-stick was concerned just as much as the electron. If we remember the second object at all we forget that it is a physical object; for us it is not a yard-stick, but just a yard.

Primarily we say *yard* rather than *yard-stick* because a great many equivalent substitutes for the yard-stick are possible. But we do not generally think of a yard as a general name for one of a large variety of physical objects or systems; we do not think of it as an object at all. I grant that another physical object may be an equivalent substitute for a yard-stick, but I do not grant that a de-materialised yard is an equivalent substitute for a yard-stick. When the quantum physicist employs a standard of length in his theory, he does not treat it as an object; if he did, he would according to the principles of his theory have to assign a wave function to it, as he does to the other objects concerned in the phenomena. In my view he is wrong. Either he is using the standard length as a substitute for the second body concerned in the observed relation of size, in which case he ought to attribute to it a wave function, so that he can bring it into his equations in the same way that the second body would have been brought in; or he is treating size as though it were not an observable relation between one physical object and another, and the lengths referred to in his formulae are not the lengths which we try to observe.*

We have to recognise then that what are called the properties of an electron are the combined properties or relations of the electron and some other physical system which constitutes a comparison object. For an electron by itself has no properties. If it were absolutely alone, there would be nothing whatever to be said about it—not even that it was an electron. And we must not be misled by the fact that in current quantum theory the comparison object is replaced by an abstraction, e.g. a metre, which does not enter into the equations in the way that an observable comparison object would do; for that is a point on which current quantum theory is clearly at fault.†

The progress of science depends on analysing our experience into its simplest elements. An object of familiar experience such as a table is found to be highly complex, so we analyse it into molecules; the molecules are found to be complex and are analysed into atoms; the atoms are found to be complex and are analysed into protons and electrons. In the pursuit of simplification we reach smaller and smaller entities until, so far as we can tell, we arrive at the limit in the electron and proton. But what meanwhile is happening to the second object concerned in the experience—the comparison object? Here our aim must be to substitute something more universal. We are dissatisfied with the yard-stick because it is clearly too local and specialised a system. To substitute an abstract yard is, as we have seen, a false step.

---

* I do not mean that they have not the same numerical value; the quantum physicist secures that by empirical adjustment. But they are quantities of different nature, and the point of practical importance is that they have a different type of probability distribution.

† The ordinary current theory is not relativistic and does not profess to be the final form. The point, however, seems to have generally been overlooked by those who are attempting to formulate a fully relativistic quantum theory. We shall refer to it again, p. 245.

We must continue to use a physical system for comparison—idealised, if you like, but not to the extent of having a different relationship to human experience from that which a physical object has. We may use a system in which a yard (or any definite number of yards) figures as a characteristic, but not a disembodied yard. In the search for universality we pass from the earth to the sun, to the "mean of the stars", to the galaxy, and finally to the most universal of all systems, namely the universe itself. In this last system a definite number of yards figures as a characteristic which is called the radius of curvature of space-time; it is thus able to serve as a comparison object for size.

The end of our pursuit of simplicity is to reach as primary object the electron (or proton), and as comparison object the universe. It is to this combination that the simplest assertions refer, and the fundamental equations of physics in their simplest form apply.

In present-day physics the most fundamental equation is the wave equation of an electron. It is usually supposed to describe the electron alone; but we have seen that that would be nonsense—there is nothing to describe. It describes the relation of the electron to a physical comparison object or system; and although the comparison system is not mentioned, we can easily see that it must be the universe—not quite the actual universe, but the universe idealised by smoothing out all gravitational and electromagnetic fields. For if a more local comparison object were involved, wave mechanics would by its own principles employ a more complicated equation with a double wave function to exhibit the observable relations involving the electron and that object.

Since the equation refers to conditions in which there is no gravitational field, the implied comparison universe is equally undisturbed. It must be remembered, however, that the wave equation has been found empirically from observations made in the actual universe; the comparison object is not just any universe, containing as much or as little matter as we like to imagine. The smoothing out of gravitational fields is just the same idealisation as is used in Lemaître's model of the expanding spherical universe; the stars and galaxies are smoothed out into a uniform distribution of matter; but the general dimensions are not tampered with. We may say briefly that in the wave equation the electron is referred to the Lemaître spherical universe as comparison object.

Thus the "wave equation of the electron" is an equation which straddles the whole of physics and describes the relation of the electron to the universe. If we invert the relation of the electron to the universe, we obtain the relation of the universe to the electron. We have only to take this equation describing the electron with the universe as comparison object, and view it, as it were, through the wrong end of the telescope, to obtain the equation describing the universe with the electron as comparison object. In describing the behaviour —in particular, the expansion—of the universe, the electron has virtually been our comparison standard; for the ordinary small-scale standards of length are constantly related to the electron. So in this way we arrive at an equation for the behaviour of the universe which (if the whole scheme of physics is consistent) must be equivalent to that given by relativity theory as developed by Friedman and Lemaître; but instead of involving a cosmical constant of undetermined value, all its coefficients are definitely known; for they are taken from the wave equation of the electron of which it is another aspect. By comparison we can accordingly find the value of the cosmical constant.

The procedure is not so simple as it sounds; but the difficulty is mainly that, before it can be applied, it is necessary to remove the fault (to which I have referred) in the existing quantum theory. Thus, although it is fairly simple in itself.

it appears as the last step in a rather difficult investigation most of which is not directly concerned with the cosmical constant.

Work on these lines has convinced me that the subject of the expanding universe is not just an interesting side-track, but is on the main route of the future development of physics. It will have a practical importance in astronomy also; for if the value of the nebular recession calculated from the ordinary laboratory constants agrees with that found by astronomical observation, it will check the accepted scale of distances of the nebulae, which is at present somewhat doubtful. I do not wish to gloss over the fragmentary state of our present knowledge; but the subject of the expanding universe seems to me to deserve prominence as one that it is of the utmost importance to continue investigating.

CHAPTER XI

THE CONSTANTS OF NATURE

Numero deus impare gaudet.*   VIRGIL, *Eclogues*.

I

In experimental science great care is lavished on the measurement of certain fundamental constants. The velocity of light, the mass of a hydrogen atom, the absolute zero of temperature, the mechanical equivalent of heat, the constant of gravitation, Planck's quantum of action—these and many others are key-numbers for the physicist. I do not think we can define a "natural constant" more particularly than as a name given to a measured quantity which is continually being used or referred to. An astronomer would rank among his natural constants the solar parallax and the constant of precession; but the importance of these is entirely local, depending on the planet which happens to be our home. Other well-known constants relate to special substances—the grating space of calcite, the wave-length of cadmium light, the specific heat of water. The selection of the first two substances depends on their practical suitability as standards; and even the choice of water might seem arbitrary to an arid Martian. But there are some constants which are evidently of the most universal significance and contain among them the scale of all natural structure; it is these which we shall here study.

But it is not easy to find a measurement which we can regard as of really universal significance. Hydrogen is the simplest of the elements; it is No. 1 in the ranking list; and we can scarcely conceive a more fundamental measurement than the determination of the diameter of the hydrogen atom

* God delights in odd numbers.

or of any equivalent linear dimension which prescribes the scale of its construction. But the difficulty is that we must measure it in terms of some standard. If report is to be trusted King Henry I, about the year 1120, fixed the yard by stretching out his arm. King David of Scotland (c. 1150) more democratically ordained that the inch should be the mean measure of the thumbs of three men "an merkle man, an man of measurable stature, and an lytell man", the thumbs being measured at the root of the nail. The metre less picturesquely embodies the mistakes of the early geodesists. Thus the result of all our careful measurement is to determine, for example, how many hydrogen atoms go to the length of King Henry's arm or to the thumbs of three Scotchmen. That does not carry us very deeply into the mysteries of Nature.

If we can measure some other length which is equally fundamental in Nature—be it the radius of the electron or of the universe, or one of those linear constants that can be compounded out of other kinds of measurement—the yard and the metre having served their purpose as intermediaries may be eliminated; and the ratio of this other length to the diameter of the hydrogen atom is a much more significant quantity. Nature is thereby measured with her own gauge. The significance of our physical constants thus lies in the purely numerical ratios which are contained in them.

The following are generally regarded as the primitive constants of physics:

e, the charge of an electron,
m, the mass of an electron,
M, the mass of a proton,
h, Planck's constant,
c, the velocity of light,
G, the constant of gravitation,
λ, the cosmical constant.

The idea is that we ought to be able to calculate out of these every other constant displayed in natural phenomena. Of course, we cannot always actually make the calculation. It may be too intricate, or we may not yet have ascertained all the rules of calculation. For example, if we want to know the wave-length of the D line of sodium, the only way at present is to measure it; but no one doubts that there exists a definite way of calculating it from the specification of a sodium atom as a nucleus surrounded by eleven electrons, employing only the seven constants listed above. I exclude, of course, constants which refer to aggregations of matter that are of only local importance, such as the dimensions of the earth or the gold equivalent of the dollar; these are obviously not deducible from the general laws of Nature.

The list of seven primitive constants is considered to be complete so far as our present knowledge extends, except that some physicists may be of opinion that an additional constant is concerned in the structure of atomic nuclei. The list can be modified by substituting equivalent constants or combinations of constants, but that does not alter anything essential. The constant h nearly always appears in theory in the combination $h/2\pi$, and it would have been better if $h/2\pi$ had been chosen as the quantum constant in the first instance. It is a common practice to denote $h/2\pi$ by $\hbar$ and regard $\hbar$ as the fundamental constant.

We may thus look on the universe as a symphony played on seven primitive constants as music is played on the seven notes of the scale. But this leaves us still under the shadow of King Henry I and his rivals; and our next step must be to free ourselves. The seven constants depend on three arbitrary units of length, of time and of mass; we describe them as having *dimensions* in length, time and mass. By forming suitable combinations of them we can replace the original seven constants by seven others, of which three are respectively a length, a time and a mass, and four are dimensionless quantities or pure numbers. Thus eliminating all reference to our artificial standards, we arrive at four pure numbers or ratios contained in the natural structure of the universe; these are in the truest sense constants of Nature, and it is on them that our interest is centred. I will give them here symbolically:

(A) $\dfrac{M}{m}$, (B) $\dfrac{hc}{2\pi e^2}$, (C) $\dfrac{e^2}{GMm}$, (D) $\dfrac{2\pi c}{h}\sqrt{\left(\dfrac{Mm}{\lambda}\right)}$.

(A) is the *mass-ratio* of the proton and electron. Its observational value is in the neighbourhood of 1840.

(B) is the *fine-structure constant*. Its observational value is about 137.

(C) is the ratio of the electrical force between an electron and proton to the gravitational force between them, if the usual laws of electrical and gravitational attraction are applicable on so small a scale. Its observational value is $2 \cdot 3 \cdot 10^{39}$.

(D) is the ratio of the natural radius of curvature of space-time to the wave-length of a mean Schrödinger wave (i.e. a geometrical mean between the wave-length associated with an electron and the wave-length associated with a proton in wave mechanics). Its value, depending on the observed recession of the spiral nebulae,* is about $1 \cdot 2 \cdot 10^{39}$.

The four numbers could be combined in various ways to give other pure numbers; but the four combinations here given furnish a natural and complete specification, such as we might provide if tenders were being invited for constructing a universe.

If we do not go beyond what is generally regarded as accepted theory, these are four ultimate constants. In that case their values have to be found by measurement; they are what they are by the whim of Nature. Ideally they are all that need be measured; and when he has supplied us with these numbers, the experimental physicist might retire and

* Taken to be 500 km. per sec. per megaparsec.

leave all the rest of physics to the mathematician. But theory is advancing, and we are beginning to ask, Are these four constants irreducible, or will a further unification of physics show that some or all of them can be dispensed with? Could they have been different from what they actually are? I am not here raising the metaphysical question whether a universe planned like ours is the only type conceivable. Granting the paraphernalia to be used, and—to save argument—granting that the number of dimensions of space-time is fixed as four, the question arises whether the above ratios can be assigned arbitrarily or whether they are inevitable. In the former case we can only learn their values by measurement; in the latter case it is possible to find them by theory.

The unification of different branches of science reduces the number of fundamental constants; and indeed the elimination of superfluous constants is the essence of the unification. We may find that experimenters in different branches of physics are measuring the same constant under different names. Eighty years ago a fundamental constant in electricity was the ratio of the electrostatic and electromagnetic units of charge; a fundamental constant in optics was the velocity of light. With the unification of electricity and optics in Maxwell's electromagnetic theory of light, these were found to be identical. More usually the constant presents itself in the two subjects in a slightly different form; thus the experimenters in one branch may have fixed their attention on the radius of a circle and those in the other branch on the circumference of the same circle. Until this is recognised by the unification of the theories, the ratio of their constants will be listed along with the other natural ratios (A), (B), (C), (D); it is superfluous when it is seen to be our old friend 2π and therefore a matter for calculation rather than measurement. The essential identity of two constants can be disguised by other factors; for example, "Stefan's constant of radiation" contains the factor $1\cdot 0823\ldots$ or $1 + (\frac{1}{2})^4 + (\frac{1}{3})^4 + (\frac{1}{4})^4 + \ldots$ ad inf.

It suggests itself that our constants (C) and (D) may have some definite theoretical relation of this kind. In that case, as soon as the theory is forthcoming, it will become unnecessary to measure more than one of them except, of course, as a useful check on the theory.

I think the opinion now widely prevails that the constants (A), (B), (C), (D) are not arbitrary but will ultimately be found to have a theoretical explanation; though I have also heard the contrary view strongly expressed. My last five years have been mainly preoccupied with this problem, and for my own part I say unhesitatingly that in the structure of the universe as known to present-day physics there is at most one arbitrary constant, viz. the large number which is the basis of (C) and (D). Our present recognition of four constants instead of one merely indicates the amount of unification of theory which still remains to be accomplished. It may be that the one remaining constant is not arbitrary; but of that I have no knowledge.

II

Let us begin with the fine-structure constant. The name has reference to the structure of spectral lines, but the constant occurs more widely; just as the velocity of light occurs in many problems unconnected with optics. The fine-structure constant is really the ratio of two natural units or atoms of action. The physical quantity rather inappropriately called "action" first became prominent in mechanics through Hamilton's Principle of Least Action; it attained further importance in relativity theory because it is one of the very few absolute quantities (invariants) in physics; and it reached its zenith in quantum theory because the quantum constant itself is a unit of action.

We obtain action when we multiply energy by time; it is *so much* energy for *so much* time. One way of obtaining a definite natural quantity of action is found by considering two elementary particles—two electrons or two protons. They have a mutual electrostatic energy due to the repulsion between them; the quantity of time naturally associated with them is the time that light (or, more appropriately, electromagnetic waves) would take to go from one to the other. Multiplying the energy by this time we obtain the action naturally associated with the pair of particles. It is always the same, whether the two particles are close together or wide apart. If they are wide apart, the energy is small but the light-time is correspondingly increased. In symbols, if r is the distance apart, the energy is e^2/r and the time is r/c, so that the product is e^2/c.

In the study of radiation another natural unit of action appears, namely $h/2\pi$. As we have already mentioned, the modern theory takes this as the unit rather than the unit h originally chosen by Planck. Thus we have two natural units of action, the one arising in electron theory and the other in radiation theory; the second is found to be approximately 137 times the first. In symbols their ratio is $h/2\pi$ divided by e^2/c, or $hc/2\pi e^2$, which is the constant (B) listed above.

So in all problems involving particles and radiation—matter and aether—we have to do with two systems constructed on a different scale, built out of different-sized atoms of action. The current theory makes no attempt to explore the significance of this difference of scale; it simply accepts it as an empirical fact and introduces an empirical fine-structure constant to allow for it. But this can scarcely be the final limit to progress. We are challenged to find a unified theory of electric particles and radiation in which the electrostatic type of action and the quantum type of action are traced to their source. We shall then be able to understand why one belongs to an atom 137 times the atom of the other, and indeed to foresee this ratio as clearly as we foresee that the circumference of a circle will be 3·14159.. times its diameter. Towards this development there is one very suggestive clue.

If we use the natural unit appropriate to the theory of electric particles (the procedure which I have described as eliminating Henry I) the mutual electrical energy, known as Coulomb energy, of two electrons or two protons distant r apart is

$$\frac{1}{r}.$$

If on the other hand we use the 137 times larger unit appropriate to the theory of radiation, the Coulomb energy is

$$\frac{1}{137} \cdot \frac{1}{r}.$$

So much for the amount of the energy. But it is a common practice to indicate physical entities by symbols which specify not only their magnitude but the direction in space or in combined space-time with which they are associated. Hence, following a method due to Dirac, it is customary to attach to the energy of two charges an additional symbol which does not affect the amount of the energy but indicates—it is difficult to say what. In more elementary problems, and in special cases, Dirac's symbols are associated with direction, and that is the notion of them that we have to start with; but the notion is extended to include anything that can make a difference to the way in which one numerical characteristic of a system can be associated with another, and it has become more or less divorced from any kind of graphical representation.* In the present case the extra symbols to be attached to the energy $1/137r$ are of the form EF, E referring to one particle and F to the other.

The attachment of such a symbol to the energy is part of the current quantum theory, which handles the energy without attempting to investigate its origin; but I would

* Some account of these symbols is given in the next chapter.

now raise the question, How many of these symbols EF have we to choose from? The answer is 136. There are 16 varieties of E and 16 varieties of F. That would give 16^2, or 256, varieties of EF. But that assumes that we can tell which of the particles has the E and which has the F. The distribution is not made that way. It is a feature of quantum theory that the particles are so much alike that we can never tell which is which; and we shall see later that this indistinguishability is actually the source of the energy that we are here studying, so that we must not ignore it here. We have then to make one of 16 possible presents to one particle and one of 16 possible similar presents to the other; but the particles are communists, not believing in private ownership, and it makes no difference which present has gone to which particle. There are 16 ways in which the commune can receive two like presents and 120 in which it can receive two unlike presents, making 136 in all. The 120 combinations of unlike presents would be duplicated if we distinguished the recipients, thus raising the number to 256.

Thus, as the energy is usually written, we have in the numerator one of 136 possible independent symbols, and in the denominator a number which is nearly 136. This surely is too striking a clue to be neglected. Let us leave aside for the moment the question of 136 or 137. The quantum physicist has in his equation an EF-symbol and he wonders why it has only $\frac{1}{136}$ of a quantum sticking to it. Is it unreasonable to suggest that the fact that EF is one of a gang of 136 may have something to do with it? Apparently the majority of quantum physicists think that it is. But for my own part the clue seems to me good enough to follow up.

But, you may say, the fraction is really $\frac{1}{137}$, not $\frac{1}{136}$. I think if we can account for $\frac{136}{137}$ of the quantum, the remaining $\frac{1}{137}$ will not be long in turning up. There is a saying, "One spoonful for each person and one for the pot".

III

We are dealing with the interaction between two elementary particles; for definiteness say, two electrons. In classical physics an interaction means a difference in the behaviour of one particle due to the presence of the other. But in quantum theory we study probability distributions and consider probable behaviour. Thus interaction means a difference in the probable behaviour of the particle. The probabilities associated with electron A, which describe its chances of occupying various positions and having various velocities, are modified by the presence of electron B.

There are two ways of treating such a change of probability. In quantum theory the actual probability is expressed as the product of an initial or a priori probability and a modifying factor. This corresponds to the Exclusion Method which we discussed earlier (p. 120). The initial probability is the frequency in an initial class, and the actual probability is the frequency in the class after it has been cut down by excluding those members which are inconsistent with the added information (supplied by observation) relating to the particular problem in hand. The change of actual probability due to the presence of electron B can be incorporated in either of the two factors—in the initial probability or in the modifying factor. It makes no difference in the long run; but the two methods correspond to somewhat different points of view. We may regard the information that electron B is present as special information relating to the particular problem under consideration; by its repulsion it keeps electron A away from it; and the probability distribution of electron A, which (but for this information) we should have anticipated, is modified accordingly. This point of view incorporates the interaction in the modifying factor. But we may also regard the fact that more than one electron is present, not as special information, but as inseparable from the statement of the problem. We regard the problem as being from the start a problem of two electrons, not as a problem of one electron into which a second electron has been introduced as an afterthought. The interaction is then incorporated in the initial probability; for, since the interaction is always present, there would be no point in considering a distribution which omits it. The fact that there *is* interaction shows that there is some kind of complication—that we do not arrive at the initial probability distribution of a crowd of particles simply by combining the probability distributions of single particles as though they were independent. For this reason a crowd of electric particles is said to obey a "new statistics", viz. Fermi-Dirac statistics, in contrast to "classical statistics" which represent the crude result of combining the probabilities independently.

Both these outlooks occur in current quantum theory and are often mixed incongruously. When a few particles are considered, the interaction is described by the Coulomb forces of repulsion or attraction and is then incorporated in the modifying factor. When many particles are considered, we use Fermi-Dirac statistics and the interaction is incorporated in the initial probability distribution. Sometimes—for greater security—it seems to get put in both ways!

Probably at this point the expert reader will object that the effects of electrical forces between the electrons are not included in Fermi-Dirac statistics. If they are not included, they ought to be. In what sense are Fermi-Dirac statistics true if they are not obeyed in Nature? It is not a question of approximation. Fermi-Dirac statistics are not intended to describe the limiting distribution when the particles are so far apart that the Coulomb forces between them are negligible. Fermi-Dirac statistics only become important when the particles are crowded together in the degenerate state of matter (p. 157); and when the Coulomb forces are negligible, the difference between Fermi-Dirac statistics and classical statistics is negligible. Why should we attach so much importance to the way in which Fermi and Dirac expected a crowd of electric particles to behave, if their expectation is not fulfilled? Fermi and Dirac may not have reached the exact expression of the interaction, but I do not think they have bluffed us by propounding a probability distribution which is *never even approximately* employed by Nature.

However that may be, we see that current theory treats interaction in a piecemeal way, partly by changing the basis of statistics and partly by introducing forces of attraction and repulsion; and so it arbitrarily divides into two compartments a subject which is really one. Both Coulomb forces and Fermi-Dirac statistics describe a change of the actual probability as compared with the classical probability for non-interacting particles; by associating one with the modifying factor and the other with the initial probability, current theory hides the fact that there is any connection between them.

It is well known that Fermi-Dirac statistics are a consequence of the fact that the particles concerned are indistinguishable from one another. But the origin of the interaction must be the same whether we express it by a new statistics or by the conception of force. If Fermi-Dirac statistics arise from the interchangeability of indistinguishable particles, so also do Coulomb forces. Coulomb energy is therefore energy of interchange. This gives a clear indication of the line to be followed in making a theoretical calculation of its value ($1/137r$). We must accomplish for two charges what Fermi and Dirac appear to have accomplished for a number of charges; that is to say, we must study in detail the way in which the probabilities connected with a system of two particles are affected by the fact that we cannot distinguish one particle from the other.

It may be useful to give an illustration showing how forces can be created by interchangeability. In astronomy the two

THE CONSTANTS OF NATURE 241

components of a double star are treated as distinguishable particles. But it happens sometimes that they are very much alike in appearance, and after a close periastron passage the observer inadvertently interchanges them. The result is an "orbit" which corresponds to a law of force unknown to Newton! If, instead of being exceptional, this were an habitual occurrence, we should be unable to verify Newton's law in double star systems. Double star astronomy, instead of being based on a law of force which, however true, could never be verified, would have to be based on a law of force appropriate to indistinguishable stars which allowed for the probability of inadvertent interchanges. The additional force would then be said (by those who recognised its origin) to correspond to energy of interchange. If you object that the force thus created is a mere fiction, I must answer that a force is not expected to be anything else but a fiction. If you are familiar with the relativity theory of gravitation you will remember that the Newtonian force is a "fiction". But whether it is a subtle fiction like gravitation, or a transparent fiction like centrifugal force, it must have its due recognition in the formulae which we use to predict observable behaviour or the probability of observable behaviour.

You will see from this example that there is nothing mystical about indistinguishability. We do not suppose that an electron knows that it will not be distinguished from other electrons, and on that account conducts itself differently. The effect of indistinguishability is that our practical problems are changed. We can imagine a higher being, who is able to identify each individual electron, continuing to treat his problems in the old way. His results are right but they do not interest us; because, not being gifted as he is, we can never find the observational data which he uses. He observes electron No. 1 at x at time t, and at x' at time t', so that he obtains the datum that it has a velocity $(x'-x)/(t'-t)$. We observe an electron at x at time t, and an electron at x' at the time t', but we do not know whether it is the same electron. Velocities of electrons are not observational data for us, for we do not know whether we are joining the observed positions rightly. We use a kind of substitute (which we can observe) called a dynamical velocity,* which may perhaps best be thought of as equivalent to the real kinematical velocity together with an allowance for an average number of mistakes of identification. But naturally these dynamical velocities do not obey the same equations as the real kinematical velocities; the difference is represented by extra terms in the equations. The mill of dynamical equations for indistinguishable particles differs from that employed for distinguishable particles because it has to grind different material.

In following up the investigation we find that the 136 symbols EF characterise different kinds of variation of a system of two electric particles or, as we commonly say, different "degrees of freedom". We have to add a 137th degree of freedom which corresponds to *interchange*. Since we cannot identify the particles, we cannot follow with certainty a particular particle through time. We may describe the system as consisting of electrons situated at P and Q, but we have to describe also the changing probability that the electron at P is electron No. 1. A diminishing probability that it is No. 1 and an increasing probability that it is No. 2, can be treated as a continuous displacement along the 137th degree of freedom which represents interchange of the two electrons. To treat the particles as distinguishable is equivalent to inserting a constraint which prevents any motion in the direction of the 137th degree of freedom; for in that case P is defined as the position of a particular particle (No. 1). The system of two distinguishable particles is accordingly limited to 136 degrees of freedom. It may be noted that

* Equal to momentum divided by mass, momentum being defined in a more general way than in elementary mechanics.

indistinguishable and distinguishable particles correspond to extreme cases, and that intermediately we can treat particles whose identification is a matter of probability, e.g. a particle which from our fragmentary information has a $\frac{2}{3}$ chance of being No. 1 and a $\frac{1}{3}$ chance of being No. 2.

The extra degree of freedom in the dynamics of indistinguishable particles explains why the quantum is divided into 137 parts instead of into 136 as we at first expected. The Coulomb energy $1/137r$ is the energy of the "motion" in the 137th degree of freedom.* That is perhaps as far as we can usefully follow the problem here. According to my own conclusions the full mathematical investigation confirms these ideas.

IV

In pursuing the fundamental constant (A), the mass-ratio, we are led into still deeper waters. By the mass of an electron we ordinarily mean the mass in the metric system, i.e. the mass in terms of a particular lump of metal called the standard kilogram. Experimental determinations of the mass are made in a variety of ways, but there are two things indispensable to every such experiment; one is the electron, and the other is the standard kilogram. The experiment may be divided into a number of steps, but together they must make a complete chain stretching from the electron to the kilo-

---

* The following additional information is for the mathematical reader. If the probability that the electron at P is No. 1 is $\cos^2 \theta$, so that the probability that it is No. 2 is $\sin^2 \theta$, the permutation variable θ is taken as the coordinate in the 137th dimension. The momentum conjugate to it is the required interaction term introduced into the dynamical equations. If the particles are indistinguishable, θ is an ignorable coordinate and the conjugate momentum has the value $1/137r$. If the particles are distinguishable, the conjugate momentum is constrained to be zero. Intermediately, when the distinction is a matter of probability, θ and its conjugate momentum have to be retained as variables in the problem.

gram lump of metal. Suppose that a physicist undertakes the complete investigation; he makes a number of observations, works out the calculations, and publishes the result; afterwards it is discovered that the standard kilogram has remained locked up in the cupboard all the time and was never used. We should conclude that there was something not quite right about his method. When we examine the current quantum theory we can scarcely avoid concluding that the theoretical physicist has been guilty of just this absent-mindedness.

But I must first explain one point. We speak of an *experimental* determination of the mass of an electron as though the experimenter were the only person concerned. But he does not put the electron in the scales and weigh it. Actually the determination is highly theoretical. The experimenter is called on to make a number of observations; but he would have no idea what observations to make, and how to get the mass of the electron out of them, if the theoretical physicist had not given him instructions. So we will discharge the experimental physicist from the problem (with a good character), and regard the currently accepted value of the mass as the discovery of the theoretical physicist who directed the operations. It is he whom I accuse of forgetting to bring in the standard kilogram.

It has happened in this way. There is generally a rather indirect chain of connection from the electron to the kilogram. The electron must be handled by quantum theory; but a big lump of metal does not require the powerful microscope of quantum theory, and classical mechanics or, for greater insight, relativity mechanics suffice. At the electron end of the experiments the quantum physicist is responsible; at the kilogram end the relativity physicist is responsible; but it seems to have been nobody's business to see that the two connect. There must always be one connecting link where we pass from a macroscopic mass to a microscopic mass; it

THE CONSTANTS OF NATURE

is this link which we need to examine, for it is here that quantum theory and relativity theory meet.

Since we have in any case one connecting link in which a microscopic mass is compared with a macroscopic mass, we may suppose for the present that the electron is being directly compared with the kilogram mass. That is to say, we have to deal with the combined probability distribution of the electron and the kilogram mass—the distribution of their positions, momenta, spins, etc. Our observational measurements determine certain features of this combined probability distribution.

It is the essence of the experiment that it determines facts about the *combined* distribution. If by any chance we can resolve it into the measurement of independent probability distributions of the electron and the kilogram separately, the experiment fails. In so far as the measurements determine properties of the probability distribution of the electron which it possesses independently of the probability distribution of the kilogram, which therefore remain when the kilogram is removed and locked up in the cupboard, they are useless for determining the mass. The theory underlying an experimental determination of mass is therefore the theory of a double probability distribution, which is described in wave mechanics by a double wave function. Double wave functions are treated at some length in books on quantum theory; but you will find no reference to double wave functions in connection with the mass of the electron. The mass is introduced as a characteristic of a single wave function referring to the probability distribution of the electron only. This, as we have seen, means that the standard kilogram is left locked up in the cupboard.

We are thus faced with a contradiction of theory and practice. In wave mechanics the mass of the electron is *defined* to be a certain characteristic of the simple wave function of the electron alone. When a so-called experimental determination of the mass has been made, it is assumed to give the value of this characteristic and is used in that way to predict spectra, collision phenomena, etc. On the other hand the experiment can only determine characteristics of the double wave function of an electron and a standard macroscopic mass. At some point therefore the theorist has inadvertently telescoped a double into a simple wave function, and we nowhere find any consideration of what is involved in this jump.

We can see how the confusion of a single and a double wave function has arisen. One of the effects of the probability waves, to which we are here referring, is to give a sort of haziness to the position of the system (cf. the "fog", p. 42). In this case we have, as it were, a two-dimensional haziness due partly to the uncertainty of position of the electron and partly to the kilogram mass. But the scale of the waves and of the resulting haziness is inversely proportional to the mass; so that in the dimension corresponding to the electron the extent of the haze is 10^{31} times greater than in the dimension corresponding to the kilogram. At first sight it seems rather pedantic to insist on the two-dimensionality of a haze which is so nearly a linear distribution. Surely it is a good enough approximation to neglect 10^{-31}! Not if 10^{-31} happens to be the quantity you are looking for, viz. the mass of the electron in terms of the standard mass. We have to insist on the two-dimensionality of the probability distribution, because a one-dimensional distribution could not include the quantity that we are measuring.

Whether or not a logical basis can be found for the replacement of the double function by a single function, we must look on it as a *fait accompli*, the consequences of which are embodied in current theory and nomenclature. It remains to inquire under what circumstances the jump can be made without ill consequences. We find that there are just two values of the mass for which the simple wave function will give the same observable results as the double wave function would have done; so that by introducing this jump the physicist (to hide his misconduct) is forced to assign to his ultimate particles one or other of these two masses. Or, to put it another way, when he separates off from the rest of the universe a portion having one or other of these two masses his rash procedure is validated; consequently he regards these masses as the only ones which can properly be separated off and treated as having an independent existence. They constitute accordingly the ultimate particles in his scheme.

For simplicity we have been speaking of a direct comparison of the mass of the electron with the kilogram; but actually the universe intervenes as an intermediary comparison body. We have seen (p. 227) that the wave equation of the electron describes the electron referred to the universe; the remainder of the chain of connection, which links the kilogram mass to the universe, is dealt with in relativity theory which exhibits the mass as a curvature of space-time. It is therefore the double wave function of the electron and the universe which has been telescoped into a single wave function. If m is the mass of the electric particle and m_0 is the mass unit furnished by the universe as comparison object, it is found that the reduction from a double to a single wave function is only possible when

$$10m^2 - 136mm_0 + m_0^2 = 0.$$

The nature of the unit m_0 need not detain us now; it is the starting point of our theory of the cosmical constant. Accepting m_0 as a definite standard unit, the quadratic equation gives two values of m, namely

$$\cdot 007357 m_0 \text{ and } 13 \cdot 593 m_0.$$

These presumably are the masses of the electron and proton, respectively. Their ratio is 1847·6, which agrees very well with the observational determinations of the mass-ratio.

V

I have called m_0 the mass unit furnished by the universe as comparison object. Since it is less than the mass of the proton it would seem odd to call it the mass of the universe. But the fact is that it is scarcely possible to define mass in a way which agrees with the practice both of quantum theory and of macroscopic theory. The terminology has grown up haphazardly and shows discrepancies when the two theories meet, and the confusion is made worse by the jump that I have described in the last section. I cannot stop here to straighten out the tangle; so I will simply say that *from a certain point of view m_0 is the mass of the universe.*

On p. 108 we found that by the Uncertainty Principle a small irreducible energy (or its equivalent, mass) is associated with any particle in curved space. We can extend this a little. I have no idea in what part of the world you, my reader, live. I might therefore describe your uncertainty of position as 4000 miles, since you must be 4000 miles in some direction from the centre of the earth. Similarly if the radius of space is R we may describe the uncertainty of position of a particle (about which we have no information) as R. If we use this particle as origin or reference point in describing the relative position of another particle, the uncertainty of position of our origin is R.

Uncertainties are diminished by taking the mean of a large number. By the theory of errors we know that if we have N independent particles, each with an uncertainty (or probable error) of position R, the uncertainty (or probable error) of position of their centre of gravity is $R/\sqrt{N}$. So if instead of using one particle as an origin of reference, we use the mean of N independent particles, we obtain an origin whose uncertainty of position is $R/\sqrt{N}$.

Let the number of particles in the universe be N. In using the universe as comparison object we smooth the distribution into a uniform spherical one corresponding to the a priori probability distribution of the particles*—i.e. the distribution assumed when we have no observational information to modify the initial probability. The particles are therefore to be treated as particles about which we have no information. The mean of these particles therefore provides a reference system or origin whose uncertainty of position is $R/\sqrt{N}$. By Heisenberg's Principle the corresponding irreducible momentum is of the order $h \div (R/\sqrt{N})$. Actually, having regard to certain details, we find it to be $h\sqrt{N}/2\pi R$. The corresponding mass is $h\sqrt{N}/2\pi Rc$, where c is the velocity of light.

The universe in its capacity as comparison object is endowed with this mass arising out of the finiteness of the space in which its N independent particles are situated. I identify this with the quantity m_0, by which we denoted the mass of the universe, or (not to shock the reader too much) the mass unit furnished by the universe as comparison object, in the last section. Thus

$$m_0 = \frac{h}{2\pi c} \cdot \frac{\sqrt{N}}{R}.$$

Although I can feel little doubt that the discussion is substantially correct, the above formula for m_0 is to some extent provisional. Among the questions that arise in the full mathematical investigation are, whether N is to be taken as the number of particles of one kind (electrons *or* protons) or the number of particles of both kinds (electrons *and* protons); whether R is the so-called Einstein radius or the de Sitter radius of the universe; whether certain averages are to be taken for three dimensions (space) or four dimensions (space-time); and so on. All these points will no doubt be settled in due course, but I am not yet ready with an unhesitating answer to them. They may introduce numerical

* See p. 132.

factors of the order 2 or 3 into the foregoing formula for m_0, but do not affect the order of magnitude.

We have already found that m_0 is $1/13\cdot593$ of the mass of a proton. Hence by the formula, employing the known values of h and c, we can obtain $R/\sqrt{N}$ (in centimetres). R/N is already known by relativity theory;* it depends on the observed value of the constant of gravitation. Combining these, we obtain R and N separately. Since the formula for m_0 is not definitive, I do not give the deduced values; but (taking the formula as it stands) they are in quite satisfactory agreement with the rather rough values obtained from the recession of the nebulae. In particular we find that N in round numbers is 10^{79}, confirming the result obtained from the nebulae (p. 221).

VI

The theoretical values of the fine-structure constant (137) and of the mass-ratio (1847·6) admit of very accurate observational test. There is no doubt that they are confirmed to within about 1 part in 500, and 1 part in 200 respectively. But it is generally considered that the precision of the observations is still greater; and they show small but significant discordances from the above values. It is to be remembered, however, that the comparison is made with so-called "observational values" which are not found by direct observation, but are calculated from observational data with formulae derived from current quantum theory; thus they contain the imperfections (if any) of current quantum theory. If the observations are regarded as com-

---

* More naturally, relativity theory gives R/M, where M is the mass of the universe (the ordinary macroscopic mass). But since we know the approximate number of electrons and protons in a gram of matter, R/N is immediately deduced.

pletely trustworthy, the discordances must lie between the current quantum theory and the new combination of relativity and quantum theory here adopted. It is scarcely fair to assume that the blame necessarily falls on the latter.

W. N. Bond has suggested that the origin of the discordances is that, in some or all of the methods of determining the mass of the electron, the adopted method of treating the observational data gives $\frac{137}{136}$ times the true mass. The constant most usually quoted is e/m in electromagnetic units; the best "observational value" of this is $1 \cdot 757 \cdot 10^7$.* On Bond's hypothesis this should be multiplied by $\frac{137}{136}$, so that the real value is $1 \cdot 770 \cdot 10^7$. It had already been found both by Bond and by Birge that $1 \cdot 770 \cdot 10^7$ is the result needed to bring about complete agreement between the observed results and the author's theoretical values of the two constants.

On theoretical grounds it seems probable that Bond's suggestion is right. We may recall that our first expectation was that the Coulomb energy of two particles would be $1/136r$. This was $\frac{137}{136}$ times too large, because we had not then noticed the 137th degree of freedom due to the indistinguishability of the particles. Bond's suggestion implies that we are not the only victims of this mistake, and that the current quantum theory in deriving (from observational data) another quantity of energy, viz. that which appears as the mass of the electron, has also obtained a result $\frac{137}{136}$ times too large. If so it is presumably due to the same cause, namely neglect of the 137th degree of freedom.

We have seen that there is nothing mystical about indistinguishability; it occasions, not difference of behaviour, but difference in what we may seek to learn about behaviour, and therefore a difference of treatment. In particular when we recognise the distinction between the particles we constrain the system to 136 instead of 137 degrees of freedom,

---

* R. T. Birge, *Nature*, April 28, 1934, p. 648.

so that we have to adopt an initial probability distributed uniformly over a closed* domain of 136 instead of 137 dimensions. This makes a difference to the average values of the various characteristics and hence to the macroscopic metrical tensor (p. 131). In other words we associate different metrics of space with the two kinds of treatment. The theoretical analysis seems to indicate that a factor $\frac{136}{137}$ (neglected in current quantum theory) is introduced by the change of metric when we equate the space occupied by the indistinguishable particles of quantum theory to the space occupied by the distinguishable parts of our measuring apparatus.

The other two numerical constants (C) and (D) cannot be found by pure theory, so far as we know; but their ratio is calculable by the theory given in Section v, so that they correspond to a single arbitrary constant in the constitution of the universe. The theoretical ratio of (C) to (D) is found to be $\sqrt{90}/\pi$, or 3·02; but this depends on our formula for m_0, which is perhaps subject to revision. It is as it stands near enough to the observed ratio having regard to the considerable observational uncertainty of (D).†

If there is only one arbitrary number in the specification of the universe it can hardly be other than the total number of particles composing it. We have called this number N. The large factor occurring in both (C) and (D) is the square root of N. Our formulae are found to give

$$(C) = \frac{\sqrt{3N}}{\pi}, \quad (D) = \sqrt{\left(\frac{N}{30}\right)}.$$

* The domain of a uniform initial probability distribution cannot have infinite volume, since that would make the probability in any finite volume zero. The domain must therefore always have the character of a spherical or hyperspherical space.

† The observed value of (C) is well determined, but the observed value of (D) which involves the recession of the nebulae is only rough.

THE CONSTANTS OF NATURE 253

For example, if we take $N = 2 \cdot 10^{79}$ (i.e. 10^{79} electrons and 10^{79} protons), we obtain

$$(C) = 2 \cdot 5 \cdot 10^{39}, \quad (D) = 0 \cdot 82 \cdot 10^{39},$$

which are near enough to the observational values on p. 232. Neither the theory nor the observational values are sufficiently reliable to warrant a closer approximation to N.

It has been remarked by Fürth that the 256th power of 2 is a little more than 10^{77}, and he has suggested that N has some connection with 2^{256}. This would mean that the number of particles in the universe has not been decided by arbitrary choice but is fixed by some inner necessity. It is an interesting conjecture, but as yet nothing has been found to encourage it.

Through the presence of $\sqrt{N}$ in the constant (C) the ratio of the gravitational forces to the electrical forces between the particles is made to depend on the total number of particles in the universe. It may be asked, Can we seriously suppose that this ratio, determinable by experiments carried out within the walls of a laboratory, is dependent on the existence of vast numbers of particles beyond the farthest galaxy yet seen? How could the abolition of all matter outside our own galaxy make any appreciable difference to an ordinary laboratory experiment such as the Cavendish experiment?

The question betrays, I think, a confusion between mathematical dependence and physical causation. I would not say that the result of the Cavendish experiment to determine the constant of gravitation is being influenced by these particles in remote parts of the universe; but the Cavendish experiment having given the result which it did, we are enabled to calculate the number of particles in remote parts of the universe assuming that the same laws hold out there. The experiment indicates the amount of curvature of space-time associated with a given number of particles. Proceeding outwards in our survey, each addition to the number of particles adds to the total bend and brings space nearer to closing up. The total number of particles is determined by the fact that the Nth particle seals the opening, and the $(N+1)$th particle is left outside—nowhere. There is no room for it in a universe whose properties accord with the Cavendish experiment. From this point of view the number of particles in the universe is a consequence of the behaviour of matter exhibited in the Cavendish experiment rather than the cause of it.

I dare say that it is somewhat of an extrapolation to regard the number N contained in the constants (C) and (D) as the actual number of particles in the universe. The theory has only been worked out for the spherical model; if there is considerable irregularity in the distribution of the galaxies, causing space to deviate widely from the spherical form, a correction may be necessary; but it is not likely that the order of magnitude will be altered. The simplest fundamental equations of physics refer to idealised uniform conditions; and it is under those conditions that the number N given by the formulae is rigorously equal to the number of particles in the whole of space.

CHAPTER XII

THE THEORY OF GROUPS

There has been a great deal of speculation in traditional philosophy which might have been avoided if the importance of structure, and the difficulty of getting behind it, had been realised. For example, it is often said that space and time are subjective, but they have objective counterparts; or that phenomena are subjective, but are caused by things in themselves, which must have differences *inter se* corresponding with the differences in the phenomena to which they give rise. Where such hypotheses are made, it is generally supposed that we can know very little about the objective counterparts. In actual fact, however, if the hypotheses as stated were correct, the objective counterparts would form a world having the same structure as the phenomenal world....In short, every proposition having a communicable significance must be true of both worlds or of neither: the only difference must lie in just that essence of individuality which always eludes words and baffles description, but which, for that very reason, is irrelevant to science.

BERTRAND RUSSELL, *Introduction to Mathematical Philosophy*, p. 61.

I

LET us suppose that a thousand years hence archaeologists are digging over the sites of the forgotten civilisation of Great Britain. They have come across the following literary fragment, which somehow escaped destruction when the abolition of libraries was decreed—

> 'Twas brillig, and the slithy toves
> Did gyre and gimble in the wabe,
> All mimsy were the borogoves
> And the mome raths outgrabe.

This is acclaimed as an important addition to the scanty remains of an interesting historical period. But even the experts are not sure what it means. It has been ascertained that the author was an Oxford mathematician; but that does not seem wholly to account for its obscurity. It is certainly descriptive of some kind of activity; but what the actors are, and what kind of actions they are performing, remain an inscrutable mystery. It would therefore seem a plausible suggestion that Mr Dodgson was expounding a theory of the physical universe.

Support for this explanation might be found in a further fragment of the same poem—

> One, two! One, two! and through and through
> The vorpal blade went snicker-snack!

"One, two! One, two!" Out of the unknown activities of unknown agents mathematical numbers emerge. The processes of the external world cannot be described in terms of familiar images; whether we describe them by words or by symbols their intrinsic nature remains unknown. But they are the vehicle of a scheme of relationship which can be described by numbers, and so give rise to those numerical measures (pointer-readings) which are the data from which all knowledge of the external universe is inferred.

Our account of the external world (when purged of the inventions of the story teller in consciousness) must necessarily be a "Jabberwocky" of unknowable actors executing unknowable actions. How in these conditions can we arrive at any knowledge at all? We must seek a knowledge which is neither of actors nor of actions, but of which the actors and actions are a vehicle. The knowledge we can acquire is knowledge of a structure or pattern contained in the actions. I think that the artist may partly understand what I mean. (Perhaps that is the explanation of the Jabberwockies that we see hung on the walls of Art exhibitions.) In mathematics we describe such knowledge as knowledge of group structure.

It does not trouble the mathematician that he has to deal with unknown things. At the outset in algebra he handles

THE THEORY OF GROUPS

unknown quantities x and y. His quantities are unknown, but he subjects them to known operations—addition, multiplication, etc. Recalling Bertrand Russell's famous definition, the mathematician never knows what he is talking about, nor whether what he is saying is true; but, we are tempted to add, at least he does know what he is doing. The last limitation would almost seem to disqualify him for treating a universe which is the theatre of unknowable actions and operations. We need a super-mathematics in which the operations are as unknown as the quantities they operate on, and a super-mathematician who does not know what he is doing when he performs these operations. Such a super-mathematics is the Theory of Groups.

The Theory of Groups is usually associated with the strictest logical treatment. I doubt whether anyone hitherto has committed the sacrilege of wrenching it away from a setting of pure mathematical rigour. But it is now becoming urgently necessary that it should be tempered to the understanding of a physicist, for the general conceptions and results are beginning to play a big part in the progress of quantum theory. Various mathematical tools have been tried for digging down to the basis of physics, and at present this tool seems more powerful than any other. So with rough argument and make-shift illustration I am going to profane the temple of rigour.

My aim, however, must be very limited. At the one end we have the phenomena of observation which are somehow conveyed to man's consciousness via the nerves in his body; at the other end we have the basal entities of physics—electrons, protons, waves, etc.—which are believed to be the root of these phenomena. In between we have theoretical physics, now almost wholly mathematical. In so far as physical theory is complete it claims to show that the properties assigned to, and thereby virtually defining, the basal entities are such as to lead inevitably to the laws which we see obeyed in the phenomena accessible to our senses. If further the properties are no more than will suffice for this purpose and are stated in the most non-committal form possible, we may take the converse point of view and say that theoretical physics has analysed the universe of observable phenomena into these basal entities. The working out of this connection is the province of the mathematician, and it is not our business to discuss it here. What I shall try to show is how mathematics first gets a grip on the basal entities whose nature and activities are essentially unknowable. We are to consider where the material for the mathematician comes from, and not to any serious extent how he manipulates the material.

This limitation may unfortunately give to the subject an appearance of triviality. We express mathematically ideas which, so far as we develop them, might just as well have been expressed non-mathematically. But that is the only way to begin. We want to see where the mathematics jumps off. As soon as the mathematics gets into its stride, it leaves the non-technical author and reader panting behind. I shall not be altogether apologetic if the reader begins to pant a little towards the end of the chapter. It is my task to show how a means of progress which begins with trivialities can work up momentum sufficient for it to become the engine of the expert. So in the last glimpse we shall have of it, we see it fast disappearing into the wilds.

II

In describing the behaviour of an atom reference is often made to the jump of an electron from one orbit to another. We have pictured the atom as consisting of a heavy central nucleus together with a number of light and nimble electrons circulating round it like the planets round the sun. In the solar system any change of the orbit of a planet takes place

THE THEORY OF GROUPS

gradually, but in the atom the electron can only change its orbit by a jump. Such jumps from one orbit to an entirely new orbit occur when an atom absorbs or emits a quantum of radiation (p. 37).

You must not take this picture too literally. The orbits can scarcely refer to an actual motion in space, for it is generally admitted that the ordinary conception of space breaks down in the interior of an atom; nor is there any desire nowadays to stress the suddenness or discontinuity conveyed by the word "jump". It is found also that the electron cannot be localised in the way implied by the picture. In short, the physicist draws up an elaborate plan of the atom and then proceeds critically to erase each detail in turn. What is left is the atom of modern physics!

I want to explain that if the erasure is carefully carried out, our conception of the atom need not become entirely blank. There is not enough left to form a picture; but something is left for the mathematician to work on. In explaining how this happens, I shall take some liberties by way of simplification; but if I can show you the process in a system having some distant resemblance to an actual atom, we may leave it to the mathematician to adapt the method to the more complex conditions of Nature.

For definiteness, let us suppose that there are nine main roads in the atom—nine possible orbits for the electron. Then on any occasion there are nine courses open to the electron; it may jump to any of the other eight orbits, or it may stay where it is. That reminds us of another well-known jumper—the knight in chess. He has eight possible squares to move to, or he may stay where he is. Instead of picturing the atom as containing a particle and nine roads or orbits, why should we not picture it as containing a knight and a chess-board? "You surely do not mean that literally!" Of course not; but neither does the physicist mean the particle and the orbits to be taken literally. If the picture is going to be rubbed out, is it so very important that it should be drawn one way rather than another?

It turns out that my suggestion would not do at all. However metaphorical our usual picture may be, it contains an essential truth about the behaviour of the atom which would not be preserved in the knight-chess-board picture. We have to formulate this characteristic in an abstract or mathematical way, so that when we rub out the false picture we may still have that characteristic—the something which made the orbit picture not so utterly wrong as the knight picture—to hand over to the mathematician. The distinction is this. If the electron makes two orbit jumps in succession it arrives at a state which it could have reached by a single jump; but if a knight makes two moves it arrives at a square which it could not have reached by a single move.

Now let us try to describe this difference in a regular symbolic way. We must first invent a notation for describing the different orbit jumps. The simplest way is to number the orbits from 1 to 9, and to imagine the numbers placed consecutively round a circle so that after 9 we come to 1 again. Then the jump from orbit 2 to orbit 5 will be described as moving on 3 places, and from orbit 7 to orbit 2 as moving on 4 places. We shall call the jump or operation of moving on one place P_1, of moving on two places P_2, and so on. We shall then have nine different operators P, including the stay-as-you-were or identical operator P_0.

We shall use the symbol A to denote the atom in some initial state, which we need not specify. Suppose that it undergoes the jump P_2. Then we shall call the atom in the new state $P_2 A$; that is to say, the atom in the new state is the result of performing the operation P_2 on the system described as A. If the atom makes another jump P_4, the atom in the resulting state will be described as $P_4 P_2 A$, since that denotes the result of the operation P_4 on the system described as $P_2 A$. If we do not want to mention the particular jumps, but to describe an atom which has made two jumps from the original state A, we shall call it correspondingly $P_b P_a A$; a and b stand for two of the numbers 0, 1, 2, ... 8, but we do not disclose which.

We have seen that two orbit jumps in succession give a state which could have been reached by a single jump. If the state had been reached by a single jump we should have called the atom in that state $P_c A$, where c is one of the numbers 0, 1, 2, ... 8. Thus we obtain a characteristic property of orbit jumps, viz. they are such that

$$P_b P_a A = P_c A.$$

Since it does not matter what was the initial state of the atom, and we do not pretend to know more about the atom than that it is the theatre of the operations P, we will divide the equation through by A, leaving

$$P_b P_a = P_c.$$

This division by A may be regarded as the mathematical equivalent of the rubbing out of the picture.

To treat the knight's moves similarly we may first distinguish them as directed approximately towards the points of the compass N.N.E., E.N.E., E.S.E., and so on, and denote them in this order by the operators Q_1, Q_2, ... Q_8. Q_0 will denote stay-as-you-were. Then since two knight's moves are never equivalent to one knight's move, our result will be*

$$Q_b Q_a \neq Q_c \text{ (unless } a, b \text{ or } c = 0).$$

We have to exclude $c=0$, because two moves might bring the knight back where it was originally.

Let us spend a few moments contemplating this first result of our activities as super-mathematicians. The P's represent activities of an unknown kind occurring in an entity (called an atom) of unknown nature. It is true that we started with

* The sign $\neq$ means "is not equal to".

a definite picture of the atom with electrons jumping from orbit to orbit and showed that the equation $P_a P_b = P_c$ was true of it. But now we have erased the picture; A has disappeared from the formula. Without the picture, the operations P which we preserve are of entirely unknown nature. An ordinary mathematician would want to be doing something definite—to multiply, take square roots, differentiate, and so on. He wants a picture with numbers in it so that he can say for example that the electron has jumped to an orbit of double or n times the former radius. But we super-mathematicians have no idea what we are doing to the atom when we put the symbol P before A. We do not know whether we are extending it, or rotating it, or beautifying it. Nevertheless we have been able to express some truth or hypothesis about the activities of the atom by our equation $P_b P_a = P_c$. That our equation is not merely a truism is shown by the fact that when we start with a knight moving on a chess-board and make similar erasures we obtain just the opposite result $Q_b Q_a \neq Q_c$.

It happens that the property expressed by $P_b P_a = P_c$ is the one which has given the name to the Theory of Groups. A set of operators such that the product of any two of them always gives an operator belonging to the set is called a *Group*. Knight's moves do not form a Group. I am not going to lead you into the ramifications of the mathematical analysis of groups and subgroups. It is sufficient to say that what physics ultimately finds in the atom, or indeed in any other entity studied by physical methods, is the *structure of a set of operations*. We can describe a structure without specifying the materials used; thus the operations that compose the structure can remain unknown. Individually each operation might be anything; it is the way they interlock that concerns us. The equation $P_b P_a = P_c$ is an example of a very simple kind of interlocking.

The mode of interlocking of the operations, not their

THE THEORY OF GROUPS

nature, is responsible for those manifestations of the external universe which ultimately reach our senses. According to our present outlook this is the basal principle in the philosophy of science.

I must not mislead you into thinking that physics can derive no more than this one equation out of the atom, or indeed that this is one of the most important equations. But whatever is derived in the actual (highly difficult) study of the atom is knowledge of the same type, i.e. knowledge of the structure of a set of unknown operators.

III

A very useful kind of operator is the *selective operator*. In my schooldays a foolish riddle was current—"How do you catch lions in the desert?" Answer: "In the desert you have lots of sand and a few lions; so you take a sieve and sieve out the sand, and the lions remain". I recall it because it describes one of the most usual methods used in quantum theory for obtaining anything that we wish to study.

Let Z denote the zoo, and S_l the operation of sieving out or selecting lions; then $S_l Z = L$, where L denotes lions—or, as we might more formally say, L denotes a pure ensemble having the leonine characteristic. These pure selective operators have a rather curious mathematical property, viz.

$$S_l{}^2 = S_l \qquad \text{(A)}.$$

For $S_l{}^2$ (an abbreviation for $S_l S_l$) indicates that having selected all the lions, you repeat the operation, selecting all the lions from what you have obtained. Putting through the sieve a second time makes no difference; and in fact, repeating it n times, you have $S_l{}^n = S_l$. The property expressed by equation (A) is called *idempotency*.

Now let S_t be the operation of selecting tigers. We have

$$S_t S_l = 0 \qquad \text{(B)}.$$

For if you have first selected all the lions, and go on to select from these all the tigers, you obtain nothing.

Now suppose that the different kinds of animals in the zoo are numbered in a catalogue from 1 to n and we introduce a selective operator for each; then

$$S_1 + S_2 + S_3 + S_4 + \ldots + S_n = I \qquad \text{(C)},$$

where I is the stay-as-you-were operator. For if you sieve out each constituent in turn and add together the results, you get the mixture you started with.

A set of operators which satisfies (A), (B) and (C) is called a *spectral set*, because it analyses any aggregation into pure constituents in the same way that light is analysed by a prism or grating into the different pure colours which form the spectrum. The three equations respectively secure that the operators of a spectral set are idempotent, non-overlapping and exhaustive.

Let us compare the foregoing method of obtaining lions from the zoo with the method by which "heavy water" is obtained from ordinary water. In the decomposition of water into oxygen and hydrogen by electrolysis, the heavy water for some reason decomposes rather more slowly than the ordinary water. Consequently if we submit a large quantity of water to electrolysis, so that the greater part disappears into gas, the residue contains a comparatively high proportion of heavy water. This process of "fractionating" is a selective operation, but it is not pure selection such as we have been considering. If taking the residue we again perform the operation of electrolysis we shall still further concentrate the heavy water. A fractionating operator F is not idempotent ($F^2 \neq F$), and this distinguishes it from a pure selective operator S.

The idea of analysing things into pure constituents and of distinguishing mixtures from pure ensembles evidently plays an important part in physical conceptions of reality. But it is not very easy to define just what we mean by it. We think of a pure ensemble as consisting of a number of individuals all exactly alike. But the lions at the zoo are not exactly alike; they are only alike from a certain point of view. Are the molecules of heavy water all exactly alike? We cannot speak of their intrinsic nature, because of that we know nothing. It is their relations to, or interactions with, other objects which define their physical properties; and in an interrelated universe no two things can be exactly alike in all their relations. We can only say then that the molecules of heavy water are alike in some common characteristic. But that is not sufficient to secure that they form a pure ensemble; the molecules which form any kind of mixture are alike in one common characteristic, viz. that they are molecules.

If we have a difficulty in defining purity of things for which we have more or less concrete pictures, we find still more difficulty with regard to the more recondite quantities of physics. Nevertheless it is clear that the idea of distinguishing pure constituents from mixtures contains a germ of important truth. It is the duty of the mathematician to save that germ out of the dissolving picture; and he does this by directing attention not to the nature of what we get by the operation but to the nature of the selective operation itself. He shows that those observational effects which reach our perceptions, generally attributed to the fact that we are dealing with an assembly of like individuals, are deducible more directly from the fact that the assembly is obtainable by a kind of operation which, once performed, can be repeated any number of times without making any difference. He thus substitutes a perfectly definite mathematical property of the operator, viz. $S_i{}^2 = S_i$, for a very vaguely defined property of the result of the operation, viz. a certain kind of likeness of the individuals which together form L. He thus frees his results from various unwarranted hypotheses that may have been introduced in trying to form a picture of this property of L.

In the early days of atomic theory, the atom was defined as an indivisible particle of matter. Nowadays dividing the atom seems to be the main occupation of physicists. The definition contained an essential truth; only it was wrongly expressed. What was really meant was a property typically manifested by indivisible particles but not necessarily confined to indivisible particles. That is the way with all models and pictures and familiar descriptions; they show the property that we are interested in, but they connect it with irrelevant properties which may be erroneous and for which at any rate we have no warrant. You will see that the mathematical method here discussed is much more economical of hypothesis. It says no more about the system than that which it is actually going to embody in the formulae which yield the comparison of theoretical physics with observation. And, in so far as it can surmount the difficulties of investigation, its assertions about the physical universe are the exact systematised equivalent of the observational results on which they are based. I think it may be said that hypotheses in the older sense are banished from those parts of physical science to which the group method has been extended. The modern physicist makes mistakes, but he does not make hypotheses.

One effect of introducing selective operators is that it removes the distinction between operators and operands. In considering the "jump" operators P, we had to introduce an operand A, for them to work on. We must furnish some description of A, and A is then whatever answers to that description. Let S_a be the operation of selecting whatever answers to the description A, and let U be the universe. Then evidently $A = S_a U$; and instead of $P_b P_a A$ we can write $P_b P_a S_a U$. Thus special operands, as distinct from operators, are not required. We have a large variety of operators, some of them selective, and just one operand—the same in every formula—namely the universe.

This mathematical way of describing everything with which we deal emphasises, perhaps inadvertently, an important physical truth. Usually when we wish to consider a problem about a hydrogen atom, we take a blank sheet of paper and mark in first the proton and then the electron. That is all there is in the problem unless or until we draw something else that we suppose to be present. The atom thus presents itself as a work of creation—a creation which can be stopped at any stage. When we have created our hydrogen atom, we may or may not go on to create a universe for it to be part of. But the real hydrogen atoms that we experiment on are something selected from an always present universe, often selected or segregated experimentally, and in any case selected in our thoughts. And we are learning to recognise that a hydrogen atom would not be what it is, were it not the result of a selective operation performed on that maze of interrelatedness which we call the universe.

In Einstein's theory of relativity the observer is a man who sets out in quest of truth armed with a measuring-rod. In quantum theory he sets out armed with a sieve.

IV

I am now going to introduce a set of operations with which we can accomplish something rather more ambitious. They are performed on a set of four things which I will represent by the letters A, B, C, D. We begin with eight operations; after naming (symbolically) and describing each operation, I give the result of applying it to ABCD:

S_α. Interchange the first and second, also the third and fourth. BADC.

S_β. Interchange the first and third, also the second and fourth. CDAB.

S_γ. Interchange the first and fourth, also the second and third. DCBA.
S_δ. Stay as you were. ABCD.
D_α. Turn the third and fourth upside down. ABƆᗡ.
D_β. Turn the second and fourth upside down. AᙚCᗡ.
D_γ. Turn the second and third upside down. AᙚƆD.
D_δ. Stay as you were. ABCD.

We also use an operator denoted by the sign — which means "turn them all upside down".

We can apply two or more of these operations in succession. For example, $S_\alpha S_\beta$ means that, having applied the operation S_β which gives CDAB, we perform on the result the further operation S_α which interchanges the first and second and also the third and fourth. The result is DCBA. This is the same as the result of the single operation S_γ; consequently

$$S_\alpha S_\beta = S_\gamma.$$

Sometimes, but not always, it makes a difference which of the two operations is performed first. For example,

Taking the result of the operation D_γ, viz. AᙚƆD, and performing on it the operation S_α, we obtain ᙚADƆ.

But taking the result of the operation S_α, viz. BADC, and performing on it the operation D_γ, we obtain BᗄᗡC.

Thus the double operation $S_\alpha D_\gamma$ is not the same as $D_\gamma S_\alpha$. There is, however, a simple relation. ᙚADƆ is obtained by inverting each letter in BᗄᗡC, that is to say, by applying the operation which we denote by the sign —. Thus

$$S_\alpha D_\gamma = -D_\gamma S_\alpha.$$

Operators related in this way are said to *anticommute*. On examination we find that S_α, S_β commute, and so do D_α, D_β; so also do S_α and D_α. It is only a combination of an S and a D with different suffixes α, β, γ (but not δ) which exhibits anticommutation.

We can make up sixteen different operators of the form $S_a D_b$, where a and b stand for any of the four suffixes $\alpha, \beta, \gamma, \delta$. It is these combined operators which chiefly interest us. I will call them E-operators and denote them by $E_1, E_2, E_3, \ldots E_{16}$. They form a Group, which (as we have seen) means that the result of applying two operations of the Group in succession can equally be obtained by applying a single operation of the Group. I should, however, mention that the operation $-$ is here regarded as thrown in gratuitously.* We may not by a single operation E_c be able to get the letters into the same arrangement as that given by $E_b E_a$; but if not, we can get the same arrangement with all the letters inverted. This property of the E-operators is accordingly expressed by

$$E_a E_b = \pm E_c.$$

We now pick out *five* of the E-operators. Our selection at first sight seems a strange one, because it has no apparent connection with their constitution out of S- and D- operators. It is as follows—

$E_1 = S_\alpha D_\delta$, which gives BADC.
$E_2 = S_\delta D_\beta$ „   AᗺCᗡ.
$E_3 = S_\gamma D_\gamma$ „   DↃBA.
$E_4 = S_\alpha D_\gamma$ „   ᗺADↃ.
$E_5 = S_\gamma D_\beta$ „   ᗡↃBA.

These five are selected because they all anticommute with one another; that is to say, $E_1 E_2 = -E_2 E_1$, and so on for all the ten pairs. You can verify this by operating with the four letters, though, of course, there are mathematical dodges for verifying it more quickly. We call a set like this a *pentad*.

There are six different ways of choosing our pentad, obtained by ringing the changes on the suffixes α, β, γ. But it is not possible to find more than five E-operators each of

* To obtain a Group according to the strict definition we should have to take 32 operators, viz. the above 16, and the 16 obtained by prefixing $-$.

which anticommutes with *all* the others. That is why we have to stop at pentads.

Another important property must be noticed. You will see at once that $E_1^2=1$; for E_1 is the same as S_α, and a repetition of the interchange expressed by S_α restores the original arrangement. But consider E_5^2. In the operation E_5 we turn the second and fourth letters upside down, and then reverse the order of the letters. The letters left right way up are thereby brought into the second and fourth places, so that in repeating the operation they become turned upside down. Hence the letters come back to their original order, but are all upside down. This is equivalent to the operation $-$. So that we have $E_5^2=-1$. In this way we find that

$$E_1^2=E_2^2=E_3^2=1, \quad E_4^2=E_5^2=-1.$$

A pentad always consists of three operators whose square is 1, and two operators whose square is -1.

With regard to the symbols 1 and -1, I should explain that 1 here stands for the stay-as-you-were operator. Since that is the effect of the number 1 when it is used as an operator (a multiplier) in arithmetic, the notation is appropriate. (We have also denoted the stay-as-you-were operator by S_δ and D_δ, so that we have $S_\delta=D_\delta=1$.) Since the operator 1 makes no difference, the operators "$-$" and "-1" are the same; so we sometimes put in a 1, when $-$ by itself would look lonely. Repetition of the operation $-$ restores the original state of things; consequently $(-)^2$ or $(-1)^2$ is equal to 1. Although the symbol, as we have here defined it, has no connection with "minus", it has in this respect the same property as $-$ and -1 in algebra.

I have told you that the proper super-mathematician never knows what he is doing. We, who have been working on a lower plane, know what we have been doing. We have been busy rearranging four letters. But there is a super-mathematician within us who knows nothing about this aspect of what we have been studying. When we announce that we have found a group of sixteen operations, certain pairs of which commute and the remaining pairs anticommute, some of which are square roots of 1 and the others square roots of -1, he begins to sit up and take notice. For he can grasp this kind of structure of a group of operations, not referring to the nature of the operations but to the way they interlock. He is interested in the arrangement of the operators to form six pentads. That is his ideal of knowledge of a set of operations—knowledge of its distinctive kind of structure. A great many other properties of E-operators have been found, which I have not space to examine in detail. There are pairs of triads, such that members of the same triad all anticommute but each commutes with the three members of the opposite triad. There are anti-triads composed of three mutually commuting operators, which become anti-tetrads if we include the stay-as-you-were operator.

All this knowledge of structure can be expressed without specifying the nature of the operations. And it is through recognition of a structure of this kind that we can have knowledge of an external world which from an ordinary standpoint is essentially unknowable.

Some years ago I worked out the structure of this group of operators in connection with Dirac's theory of the electron. I afterwards learned that a great deal of what I had written was to be found in a treatise on Kummer's quartic surface. There happens to be a model of Kummer's quartic surface in my lecture-room, at which I had sometimes glanced with curiosity, wondering what it was all about. The last thing that entered my head was that I had written (somewhat belatedly) a paper on its structure. Perhaps the author of the treatise would have been equally surprised to learn that he was dealing with the behaviour of an electron. But then, you see, we super-mathematicians never do know what we are doing.

V

As the result of a game with four letters we have been able to describe a scheme of structure, which can be detached from the game and given other applications. When thus detached, we find this same structure occurring in the world of physics. One small part of the scheme shows itself in a quite elementary way, as we shall presently see; another part of it was brought to light by Dirac in his theory of the electron; by further search the whole structure is found, each part having its appropriate share in physical phenomena.

When we seek a new application for our symbolic operators E, we cannot foresee what kind of operations they will represent; they have been identified in the game, but they have to be identified afresh in the physical world. Even when we have identified them in the familiar story of consciousness, their ultimate nature remains unknown; for the nature of the activity of the external world is beyond our apprehension. Thus armed with our detached scheme of structure we approach the physical world with an open mind as to how its operations will manifest themselves in our experience.

I shall have to refer to an elementary mathematical result. Consider the square of $(2E_1+3E_2)$, that is to say the operation which is equivalent to twice performing the operation $(2E_1+3E_2)$. We have not previously mixed numbers with our operators; but no difficulty arises if we understand that in an expression of this kind 2 stands for the operation of *multiplying* by 2, 3 for the operation of multiplying by 3, as in ordinary algebra. We have

$$(2E_1+3E_2)(2E_1+3E_2) = 4E_1^2 + 6E_1 E_2 + 6E_2 E_1 + 9E_2^2.$$

We have had to attend to a point which does not arise in ordinary algebra. In algebra we should have lumped together the two middle terms and have written $12E_1 E_2$ instead of

$6E_1E_2+6E_2E_1$. But we have seen (p. 269) that the operation E_2 followed by the operation E_1 is not the same as the operation E_1 followed by the operation E_2; in fact we deliberately chose these operators so that $E_2E_1=-E_1E_2$. Consequently the two middle terms cancel one another and we are left with

$$(2E_1+3E_2)^2=4E_1{}^2+9E_2{}^2.$$

But we have also seen that $E_1{}^2=1$, $E_2{}^2=1$. Thus

$$(2E_1+3E_2)^2=4+9=13.$$

In other words $(2E_1+3E_2)$ is the square root of 13, or rather *a* square root of 13.

Suppose that you move to a position 2 yards to the right and 3 yards forward. By the theorem of Pythagoras your resultant displacement is $\sqrt{(2^2+3^2)}$ or $\sqrt{13}$ yards. It suggests itself that when the super-mathematician (not knowing what kind of operations he is referring to) says that $(2E_1+3E_2)$ is a square root of 13, he may mean the same thing as the geometer who says that a displacement 2 yards to the right and 3 yards forward is square-root-of-13 yards. Actually the geometer does not know what kind of operations he is referring to either; he only knows the familiar story teller's description of them. He can render himself independent of the imaginations of the familiar story teller by becoming a super-mathematician. He will then say:

What the familiar story teller calls displacement to the right is an operation whose intrinsic nature is unknown to me and I will denote it by E_1; what he calls displacement forward is another unknown operation which I will denote by E_2. The kind of knowledge of the properties of displacement which I have acquired by experience is contained in such statements as "a displacement 2 yards to the right and 3 yards forward is square-root-of-13 yards". In my notation this becomes "$2E_1+3E_2$ is a square-root of 13". Super-mathematics enables me to boil down these statements to the single conclusion that displacement to the right and displacement forward are two operations of the set whose group structure has been investigated in Section IV.*

Similarly we can represent a displacement of 2 units to the right, 3 units forward and 4 units upward by $(2E_1+3E_2+4E_3)$. Working out the square of this expression in the same way, the result is found to be 29, which agrees with the geometrical calculation that the resultant displacement is $\sqrt{(2^2+3^2+4^2)}=\sqrt{29}$. The secret is that the super-mathematician expresses by the *anticommutation* of his operators the property which the geometer conceives as *perpendicularity* of displacements. That is why on p. 269 we singled out a pentad of anticommuting operators, foreseeing that they would have an immediate application in describing the property of perpendicular directions without using the traditional picture of space. They express the property of perpendicularity without the picture of perpendicularity.

Thus far we have touched only the fringe of the structure of our set of sixteen E-operators. Only by entering deeply into the theory of electrons could I show the whole structure coming into evidence. But I will take you one small step farther. Suppose that you want to move 2 yards to the right, 3 yards forward, 4 yards upward, and 5 yards perpendicular to all three—in a fourth dimension. By this time you will no doubt have learned the trick, and will write down readily $(2E_1+3E_2+4E_3+5E_4)$ as the operator which symbolises this displacement. But there is a break-down. The trouble is that we have exhausted the members of the pentad whose square is 1, and have to fall back on E_4 whose square is -1 (p. 270). Consequently

$$(2E_1+3E_2+4E_3+5E_4)^2=2^2+3^2+4^2-5^2=4.$$

* He will, of course, require more than a knowledge relating to two of the operators to infer the group structure of the whole set. The immediate inference at this stage is such portion of the group structure as is revealed by the equations $E_1^2=E_2^2=1$, $E_2E_1=-E_1E_2$.

THE THEORY OF GROUPS   275

Thus our displacement is a square root of 4, whereas Pythagoras's theorem would require that it should be the square root of $2^2+3^2+4^2+5^2$, or 54. Thus there is a limitation to our representation of perpendicular directions by E-operators; it is only saved from failure in practice because in the actual world we have no occasion to consider a fourth perpendicular direction. How lucky!

It is not luck. The structure which we are here discussing is claimed to be the structure of the actual world and the key to its manifestations in our experience. The structure does not provide for a fourth dimension of space, so that there cannot be a fourth dimension in a world built in that way. Our experience confirms this as true of the actual universe.

If we wish to introduce a fourth direction of displacement we shall have to put up with a minus sign instead of a plus sign, so that it will be a displacement of a somewhat different character. It was found by Minkowski in 1908 that "later" could be regarded in this way as a fourth direction of displacement, differing only from ordinary space displacements in the fact that its square combines with a minus instead of a plus sign. Thus 2 yards to the right, 3 yards forward, 4 yards upward and 5 "yards" later* is represented by the operator $(2E_1+3E_2+4E_3+5E_4)$. We have calculated above that it is a square root of 4, so that it amounts to a displacement of 2 yards. When, as here, we consider displacement in time as well as in space, the resultant amount is called the *interval*. The value of the interval in the above problem according to Minkowski's formula is 2 yards, so that our results agree. Minkowski introduced the minus instead of the plus sign in the fourth term, regarding the change as expressing the mathematical distinction between time and space; we introduce it because we cannot help it—it is forced on us by the group structure that we are studying. Minkowski's interval

* A "yard" of time is to be interpreted as the time taken by light to travel a yard.

afterwards became the starting point of the general theory of relativity.

Thus the distinction between space and time is already foretold in the structure of the set of E-operators. Space can have only three dimensions, because no more than three operators fulfil the necessary relationship of perpendicular displacement. A fourth displacement can be added, but it has a character essentially different from a space displacement. Calling it a time displacement, the properties of its associated operator secure that the relation of a time displacement to a space displacement shall be precisely that postulated in the theory of relativity.

I do not suggest that the distinction between the fourth dimension and the other three is something that we might have predicted entirely by a priori reasoning. We had no reason to expect a priori that a scheme of structure which we found in a game with letters would have any importance in the physical universe. The agreement is only impressive if we have independent reason to believe that the world-structure is based on this particular group of operators. We must recall therefore that the E-operators were first found to be necessary to physics in Dirac's wave equation of an electron. Dirac's great achievement in introducing this structure was that he thereby made manifest a recondite property of the electron, observationally important, which is commonly known as its "spin". That is a problem which seems as far removed as possible from the origin of the distinction of space and time. We may say that although the distinction of space and time cannot be predicted for a universe of unknown nature, it can be predicted for a universe whose elementary particles are of the character described in modern wave mechanics.

It only remains to add that the sixteen E-operators are those referred to in the previous chapter (p. 236). We there use a double set; e.g. if E_1 signifies the operation of displacing electron No. 1 to the right, F_1 denotes the operation of displacing electron No. 2 to the right. I have already explained the way in which the mystic number 136 arises in this double set. The double set of operators is not confined to the particular problem of two particles. It has universal application owing to the fact that we can only observe relations; therefore our standard equipment consists of two sets of operators, one for each end of the relation. For that reason the number 136 appears again in a different connection on p. 247.

There are two ways in which the number 136 is involved in a double-set of E-operators. I have explained one way; I will now explain the other way, which is I believe more directly the origin of its occurrence in the fundamental constants of Nature. We have distinguished operators such as E_1 whose square is 1 from those such as E_4 whose square is -1. We may call the first "space-like" and the second "time-like", since that is the way in which the distinction has appeared so far as we have investigated it in the foregoing discussion. Classifying all the 16 E-operators in this way, we find that 10 of them are space-like and 6 are time-like. Classifying similarly the 256 EF-operators, we find at once that $10^2 + 6^2$ or 136 are space-like (square=1), and $(10 \times 6) + (6 \times 10)$ or 120 are time-like (square=-1). I think that when 136 occurs in the constants of physics it generally refers to the 136 space-like double operators; for the space-like operators determine the number of dimensions of the domain over which the total probability of a system is distributed. But to pursue these questions would take us too deeply into the theory.

CHAPTER XIII

CRITICISMS AND CONTROVERSIES

Codlin's the friend, not Short. Short's very well as far as he goes, but the real friend is Codlin—not Short.
DICKENS, *The Old Curiosity Shop.*

I

IN the early days of the theory of relativity one of the most frequent questions asked by my correspondents was, Is the FitzGerald contraction real or apparent? Is it really true that a rapidly moving rod becomes shortened in the direction of its motion? The answer which I gave in *The Nature of the Physical World* (pp. 32–34) is too long to quote here; but, having pointed out with an example that we often draw a distinction between things which are "true" and things which are "really true", I explained that on the same principles the contraction of the moving rod would be described as *true* but not *really true*.

It is interesting to note the reaction of a not unfriendly philosophical reviewer* towards this effort to explain:

Surely it is simpler to say straight out that distance between two particles is not a dual relation but a triple relation into which a frame of reference enters and that the only valuable dual relation that can be extracted is in the cases when the two particles are relatively at rest and are permitted to fix the frame of reference as one in which they are at rest.

Very much simpler—for the author, at any rate.

Non-technical books are very often a target for criticism simply because they are non-technical. I have quoted a very innocent example of such criticism. But one does not so

* *Mind*, vol. 38, p. 413 (1929).

easily excuse the critic who imputes scientific laxity on no better grounds. Statements are described as careless because they are not hedged about with safeguards like legal documents. Explanations are treated as definitions. The author is convicted of not saying what he means. Of course he does not say exactly what he means; in ordinary speech one seldom does. The understanding between a non-technical writer and his reader is that he shall talk more or less like a human being and not like an Act of Parliament.

I take it that the aim of such books must be to convey exact thought in inexact language. The author has abjured the technical terms and mathematical symbols which are the recognised means of securing exact expression, and he is thrown back on more indirect methods of awakening in the mind of the reader the thought which he wishes to convey. He will not always succeed. He can never succeed without the cooperation of the reader.

A correspondent has pointed out to me that in various places in *The Nature of the Physical World* the word "space" occurs with four different meanings. I think he expected me to feel penitent. But the word *has* these meanings; and inasmuch as my correspondent was by no means diffident in telling me what I really meant in each place, I inferred that in this instance my attempt to convey exact thought in inexact language had succeeded.

It is not a question of stepping down from the austere altitude of scientific contemplation to a plane of greater laxity. To free our results from pedantries of expression, and to obtain an insight in which the less essential complications do not obtrude, is as necessary in research as in public exposition. We strive to reduce what we have ascertained to an exact formulation, but we do not leave it buried in its formal expression. We are continually drawing it out from its retreat to turn it over in our minds and make use of it for further progress; and it is in this handling of the truth that the rigour of scientific thought especially displays itself. Rigour is a much misused term, and not only in expository writing but in original scientific investigations it is too frequently another name for lack of a sense of proportion.

II

In reading the various discussions which have arisen out of the philosophical position that I have taken up in my earlier writings, it has seemed to me that the most urgent point of controversy is the deadlock referred to in Chapter I concerning my remark "Mind is the first and most direct thing in our experience; all else is remote inference". The typical philosopher and the typical scientist seem to have taken up irreconcilable positions.

First let me summarise my own view—which is, I think, acceptable to most scientific men who have reflected at all on the subject. The experience of each individual is primarily the changing content of his consciousness. It is a succession of tableaux accompanied by sensory feelings of various kinds, memories, abstract thoughts, emotions, etc. Even to the least reflective of us this complex activity presents a problem; we want to find the associations of the various elements in this experience, and to make out what it is all about. But for the scientist at least the nature of the inquiry is very largely determined by the fact that individuals are able to communicate part of their experience to one another. When you tell me your experience the sound of your voice is part of my experience; but for the purposes of the problem it is not *my* awareness of a sound that I utilise, but *your* awareness of something else. I treat it as an admission to an experience—a content of consciousness—which is not my own. By this step the problem is enlarged; it is no longer a matter of determining the interrelatedness of elements all contained in one consciousness, but the interrelatedness of elements in many different consciousnesses; it therefore requires the conception of a theatre of activity external to the individual consciousness. Physical science gives us a picture, or more strictly a symbolic formulation, of such an external theatre of activity interacting with each individual consciousness. If we accept the scientific solution, and in particular the scientific account of the nerve mechanism of the body, the connection between the objects inferred to exist in this external world and the sensations experienced in consciousness is evidently remote and indirect.

It is very difficult to see what the philosopher is after when he challenges this. It is important to discover whether it is simply the common kind of misunderstanding which arises when two people do not talk the same language, or whether the philosopher really intends to reject the scientific account of the origin of our perceptions so far as that origin lies outside consciousness itself.

Two philosophical writers have entered into this question in some detail with special reference to my own writings, namely Prof. W. T. Stace* and Mr C. E. M. Joad.† I will deal with them separately since they take up rather different positions. I understand that between them they represent a considerable body of opinion among philosophers of the realist school.

I might define my typical opponent as the man who believes in the existence outside the mind of "an actual apple with an actual taste in it". I do not object to an actual apple external to the mind, and I am willing to be convinced as to the existence of an actual taste (as distinct from the physical interaction between the molecules of the apple and those of a particular tongue) external to the mind. Where the philosopher seems to fly against the plain teaching of science is in locating the actual taste in the actual apple. It is better to avoid words such as taste, colour, sound, which are used

* "Sir Arthur Eddington and the Physical World", *Philosophy*, vol. 9, p. 40 (1934).   † *Philosophical Aspects of Modern Science* (1932).

confusingly both for the sensation and the indirect cause of the sensation. It might be clearer therefore if I described the philosophers in question as those who believe in "an actual dentist's drill with an actual pain in it", which is, I suppose, an obvious corollary. But I am afraid of saddling them with opinions which I have not seen explicitly admitted.

The kind of datum from which scientist and philosopher alike must start is exemplified by I-perceive-the-taste-of-an-apple. I use this string of words to indicate to you a particular kind of awareness; it is the awareness, not the description nor the analysis implied in the description, which constitutes the datum. Another datum may be he-perceives-the-sound-of-a-bell. It is agreed that although I have become possessed of this second datum in an indirect way it ranks equally with the first which is immediately furnished by my consciousness. The recognition that sense-data may have different subjects ("I" and "he") is the first step in their analysis. It suggests itself for trial that they may be treated as subject-object relations. We can form a kind of equation—

Datum *minus* Subject *minus* a constant relation ("perceives") = Object.

But if this is to carry us any further we must suppose that some at least of the objects are capable of association with different subjects. We then have a *communal object* the-sound-of-a-bell which is common to the data I-perceive-the-sound-of-a-bell and he-perceives-the-sound-of-a-bell; just as the subject "I" can be common to a number of data. The view that the objects of sense-data are communal objects, capable of perception by more than one subject, is a hypothesis; it implies that the gamut of sensations of one individual can in some way be identified with those of another individual.

The data are evidently mental; they are an awareness—a content of the consciousness of myself or of so-and-so. The communal objects, if they exist, are not in any one consciousness; nor are they to be identified with the objects of the physical world. If it is necessary to locate them anywhere it must be in some third territory. Prof. Stace suggests the following as a view to which the physicist ought not to object—

The view that sensory qualities are mental depends on the uncriticized dogma that there are only two realms to which they could belong, the physical and the mental. If this were so, then to prove that they are not physical would be the same as proving that they are mental. And this is what the physicist does. But the assumption on which the argument is based, the traditional common-sense division of the universe into mind and matter and nothing else is false. There is a third realm, which is neither physical nor mental, but which we may call the "neutral" realm. Sensory qualities belong to this realm and are neither physical nor mental.

It is true that I commonly use the words mind and mental to cover all that is non-physical in the same way that matter and material are frequently used to cover all physical entities. There is I think no more comprehensive term which would be generally intelligible. The paucity of language is illustrated by Prof. Stace's suggested term *neutral*. It suggests that the sensory qualities (i.e. communal objects) are neutral as between the physical and the mental realm; and it is evident from later statements that that is how Stace regards them. But from a physicist's point of view they are ultra-mental. I mean that when from our common sense-data the philosopher derives a communal object the-sound-of-a-bell, and still more when he derives (I know not how) a bell with a sound in it, he is proceeding away from the external world of physics; to reach the bell contemplated by the physicist these steps have to be retraced and we have to start again from the immediate awareness of individual minds—as I shall show later. Prof. Stace continues—

From this point of view Sir Arthur's statement, "Mind is the first and most direct thing in our experience; all else is remote inference"—which, he says, horrified the philosophers so much—would appear to have nothing horrifying in it except the apparent identification of sense-data with minds. *That*, I suppose, was what shocked the philosophers.

I do not admit that the philosopher's construct of a realm of sensory qualities (communal objects) existing outside our individual minds is any less a remote inference than the construct of a physicist from the same data. An "impersonal taste" not forming part of the content of anyone's consciousness may be a legitimate conception, but it is not a matter of immediate experience; at least I have never tasted one. I must make it clear that for me sensory data are the experiences themselves—the awareness of someone of something; whereas Stace and perhaps philosophers generally use the term sense-data for what I here call communal objects such as the-taste-of-an-apple, or even for an object supposed to be compounded of tastes, colours, shapes, etc. like the familiar apple described by the story teller in my consciousness. To call such hypothetical constructs "data" seems to me to beg the question.

Let us trace the first few steps in a scientist's inference from the data of his experience. In the first place he finds the data in which he himself plays the part of subject arranged in a time sequence. A more primitive description of a state of awareness would be I-perceive-a-taste-which-I-perceived-yesterday. We recognise a sensation not for what it is in itself but for its resemblance to a previous sensation. Attention at once passes away from the nature of the experience to the recurrency of the experience. As shown in the first chapter, it is out of the recurrencies of experience that the world of physics is inferred. The scientist's path therefore diverges immediately from that of the philosopher. The latter forms a general object or sensory quality, the-taste-of-an-apple, not in any one person's experience and therefore not associated with any particular place or time; any recurrency is lost by removing it from the time sequence of its subject. It may happen that a number of individuals (bent on keeping the doctor away) perceive the-taste-of-an-apple with diurnal recurrency. It seems to me that the philosopher who starts by treating the-taste-of-an-apple as the object of an experience common to them all and therefore dissociated from individual time sequences is unable to deal with this recurrency even though it is a recurrency of their common experience.

The point that I have to bring out is that the communal objects or sensory qualities have no bearing on physical science, since they eliminate the very part of experience with which the physicist is concerned, namely recurrency, and retain the part with which he is not concerned, namely the qualitative character of the sensations. Therefore the sensory qualities discussed by the philosophers do not form a realm intermediate between the mental and the physical world.

At first sight it seems very satisfactory that the philosopher should care for that which the physicist neglects and *vice versa*. But I feel bound to ask the question whether the philosopher starting from the same data of experience has deliberately tackled a different problem and reached a different domain of truth, or whether he has attempted the same problem of synthesis as the physicist and has lost his way at the start. When the philosopher gives externality and objectivity to particular tastes I am not convinced that his basis lies in philosophy at all; it looks to me more like second-hand physics.

The recognition of a taste of a particular quality as existing outside an individual consciousness must depend on identifying a sensation of taste by one individual with a sensation of taste by another individual. Therefore I take it that the philosopher supposes the taste sensations of different individuals to have a one-to-one correspondence, and he interprets this correspondence as signifying identity of the object—the taste—perceived. Now the scientist also places the taste sensations of different individuals in one-to-one correspondence, without however speculating as to whether corresponding sensations are identical. For him corresponding (but not necessarily identical) sensations of taste are those which arise when the tongues of different people are stimulated by like objects. Is it this correspondence which the philosopher has taken over and reinterpreted? If not, how does he define his correspondence? Prof. Stace's view seems to be that there is floating round in his third "neutral" realm a particular taste capable of being perceived either by you or by me. If the taste is one that I perceived on a specified occasion, how are we to know when you perceive the same taste? The question demands an answer; for clearly the existence of a taste common to the perception of both of us cannot be one of the "first things in our experience" if we do not even know when the experience occurs.

When a philosopher describes the taste as being the taste of an apple, it looks as though he had borrowed the scientist's criterion of correspondence. He apparently refers to the scientist's test of placing portions of an apple, i.e. a hard, round, green object with specified antecedents in the physical world, on the tongues of various individuals; and argues that because the physical conditions are similar the sensation in each mind is a perception of a common object of perception. But if that is really the way in which the philosopher discovers (or invents) the impersonal tastes which occupy his non-mental non-physical realm, these tastes, so far from being immediately known to exist, are a remote inference from our remote inferences about the physical world.

If every element of experience was utterly unlike every other element of experience there would be no subject-matter either for science or philosophy. Progress depends on recognising a common factor in two elements of ex-

CRITICISMS AND CONTROVERSIES 287

perience. The most elementary type of common factor is indicated by our recognition that a number of experiences may have the same subject; we may pass over this, since no controversy arises. After this we must study further resemblances between two elements of experience, either (*a*) which have the same subject, or (*b*) which have different subjects. Physics is based on (*a*); neo-realist philosophy seems to be based on (*b*). In the latter the common factor is supposed to be a common object of perception inhabiting the "third" realm. In the former the common factor is not identified with an object of perception. (Physical objects are not reached until a much later stage of inference.) We use instead the group-structure of the resemblances (recurrencies) which by the theory of Chapter XII can be conceived to have an existence independent of that in which it is a structure. By this step we transfer the structure of individual experience into an external domain where it can be interwoven with the structures of other individual experiences similarly transferred.

If the resemblances (*b*) were immediate data of experience—something we were directly aware of—there would be a justification for saying that the objects of perception, or sensory qualities, are less remote inferences than are physical objects. But it is obvious that the resemblances (*a*) are the only ones of which we are directly aware. Resemblances (*b*) are not only remote but very uncertain inferences from our experience, and they presuppose a detailed knowledge of physics and physiology. I cannot help thinking that your sensation when you are eating an apple resembles my sensation in eating an apple more closely than it does my sensation in hearing a bell; but I do not know how to give any logical defence for this opinion—because indeed resemblances of type (*b*) are so completely outside experience that we can form no idea of what such a resemblance would mean.

When the philosopher proceeds further to associate

the-taste-of-an-apple with an apple, he is attacking the problem of finding the association between certain perceptions of taste, sight and touch which has also been attacked by the physicist, and more especially by the physiologist. If he neglects the experimental method of finding out the nature of the association, and propounds a hypothesis of their combined existence in an object outside the mind but directly apprehended by a number of minds—a hypothesis directly contradictory to the scientific theory of the way in which the associated recurrence of sensations is determined—it is inevitable that in the eyes of a scientist he should be classed with the idle speculators.

I hope that this reply will not be looked on as an attempt to make public exposure of the hollow pretensions of realist philosophers. I am well aware that Prof. Stace cannot use all his armoury in a non-technical article. If sometimes he has not said quite what he means, I may still be to blame for not grasping his thought, seeing that I have not sufficient technical knowledge to follow the language of an exact statement. My intention has been to make clearer the case which philosophers have to meet, and to show that the new scientific philosophy is not quite the defenceless victim that some of them are apt to assume.

III

I think that the fullest criticism of my scientific philosophy is contained in Mr C. E. M. Joad's *Philosophical Aspects of Modern Science*. I must first recognise the great care with which he has presented my arguments and conclusions, distinguishing them where necessary from those of Jeans and Russell with which he also deals. Mr Joad belongs to the realist school, and therefore our controversy is to some extent the same as that discussed in the last section. But he appears to take a less extreme view than Prof. Stace. According to

CRITICISMS AND CONTROVERSIES 289

Stace "Chairs and tables and stars do really exist. They are exactly what they appear to be, coloured, spatial, resounding objects. Moreover, this familiar world is the only real world, the only world which really exists". Joad says (p. 12) "I cannot claim—nor, indeed, did I expect—that so far as the vindication of the common-sense world is concerned the attempt has met with much success. Chairs and tables do not, I fear, exist in the way in which in everyday life we suppose them to exist; but it does not follow that the physicist's analysis of them into atoms and electrons invested with secondary qualities projected upon them by the mind is therefore correct". And again (p. 11) "The neo-realists' world of correlated sense data has borne less and less resemblance to the common-sense world of physical objects which realism began by trying to preserve".

Naturally this more moderate attitude brings him very much closer to my own view:

> I should hold, then, that the researches of the scientist are, equally with the perceptions of the plain man, the moral consciousness of the good man, the sensitivity of the artist and the religious experience of the mystic, revelatory of reality. Epistemologically they stand on equal terms. Such arguments as there are for supposing that any of these forms of experience is *merely* subjective, apply also to the others; but equally if any of them gives us information about a world external to ourselves, so also do the others.

I might easily have mistaken that for an extract from my own writings.

But the closeness of the approach makes me inclined all the more to question the need for a neo-realist's world which is neither the familiar world nor the scientific world nor a wider reality containing the scientific world as part of itself. When two men's paths are at right angles we suppose that they have different but equally valid objectives; when they diverge at a narrow angle we suspect that one of them may have mistaken the way.

Mr Joad's criticisms are mainly directed to discovering inconsistencies of expressions and ideas in my writings. It is tempting to enter on a detailed defence; but it is perhaps better to confine myself to a general observation. I do not think that such discrepancies will appear so heinous to a scientist as they do to a philosopher. In science we do not expect finality. The theories described in the scientific part of this book do not form a complete and flawless system; there are incoherencies which we cannot remedy until further research gives us new light. It may well be that the scientific theory will be substantially modified in its future progress towards completion; nevertheless we feel justified in claiming that our present imperfect results embody a large measure of truth. I naturally look on scientific philosophy as subject to the same progressive advance. Undoubtedly the recent developments of physics have philosophical implications of the highest importance, and I have endeavoured to explain and elaborate them. As with the scientific advances, so the philosophical advances can be consolidated into something like a system; but it does not disturb me unduly if there are loose ends that do not quite fit into the system—glimpses of a deeper truth which we are not yet able to formulate.

The advance of scientific philosophy has come from two main sources, the relativity theory and the quantum theory. When *The Nature of the Physical World* was written the scientific conceptions of the two theories were conflicting; and although there is now no longer a definite conflict the unification is still incomplete. It is not surprising that the philosophical outlook should display traces of the same discrepancy; but if the respective philosophies of relativity theory and quantum theory are not entirely harmonious they have at all events a large common denominator.

I do not know whether it would be fair to say that the

CRITICISMS AND CONTROVERSIES 291

philosopher lays so much stress on formal consistency because he has little else by which to test the validity of his philosophy. But at any rate this does not apply to a philosophy developed on a scientific foundation. In science the unity and consistency of the system is an ideal to be reached by convergence. We are accustomed to finding different aspects of truth according to the way we approach it; we rejoice in its many-sidedness. I have been at no pains to suppress the many-sidedness of the truth which I believe is contained in the modern advance of physical science; and I therefore fall an easy victim to anyone who cares to collate passages in which I approach it from different angles.

The shallower critics have also made capital by mixing passages in which the outlook is purely scientific, or passages in which I was leading the reader on, with those giving explicit statements of my philosophical ideas. In this respect Mr Joad has been entirely fair. It is necessary, however, to call attention to one surprising lapse. According to him, I affirm that atoms and electrons are objectively real and in fact that they (with other physical entities) constitute the sole objective reality. He asserts this twice (*loc. cit.* pp. 113, 127) and on each occasion he quotes in support a sentence almost at the beginning of my book, "modern physics has assured me that my second, scientific table is the only one that is really there—wherever 'there' may be". Surely the effect of the last four words is to suggest that there is a loophole, and that the assurance may appear in a different light when we discover what meaning (if any) is to be attached to "there". I certainly do not regard the entities of the physical world as the sole objective reality. As to whether atoms and electrons are objectively real, I divide my answer into two parts. Firstly, I do not think it is very important whether or not we use a particular phrase "objectively real", which nobody seems able to define. I have tried to explain the relation of atoms and electrons to the data of human experience. I think that the reader will be inclined to call whatever has this relation to experience "real"; but if he considers that it is an insufficient qualification for reality I shall not demur. It is purely a question of definition. Secondly, since atoms and electrons are the subject of quantum theory which is still in course of development, their *scientific* status is subject to some uncertainty, and this naturally affects their philosophical status. The effect of wave mechanics (especially as developed by Dirac) is to make the separation of the subjective and objective elements in human experience more indefinite. Relativity theory revealed an unsuspected subjective element in classical physics and cleared it away; wave mechanics has revealed a further subjective element; but its procedure is to let it stay and adopt methods suitable for treating a partially subjective world. So far as I can see, we find ourselves unable to reach by physical methods a purely objective world, and it would seem to follow that all the entities of physics have the partial subjectivity of the world to which they belong—though, of course, they are not purely subjective.

Turning to some of Joad's special criticisms, he objects (p. 31) that I represent the physical world as (a) abstracted by the mind of the scientist from a more comprehensive reality, (b) as constructed by the mind from relations and relata, and (c) given as embedded in a background of reality; according to him it cannot be all three, i.e. abstracted, constructed and given. But I see no reason why all three conceptions should not be applicable. The constellation Draco is a two-dimensional appearance of stars *abstracted* from their complete spatial distribution; it is *constructed* by a mind which is seeking resemblances to mythological characters and creatures; it is *given* as embedded in a galaxy of stars. If any one of these three aspects were missing the constellation Draco would not exist. Obviously it would not exist if there were no stars; there is no association between the stars composing it other than a fanciful resemblance to a serpent; and this resemblance only exists because it is contemplated as projected on the sky instead of in three dimensions. Similarly the world described by the equations of physics is, I believe, embedded in a background external to the individual mind, and constructed by a putting together of associations to which the mind is sensitive; its abstract character is obvious.

There follow a number of criticisms (pp. 34-41) which suggest that Mr Joad has not grasped what is implied by the symbolic character of physical entities. It is as though, having said "Let x be the mass", I was supposed to be guilty of confusion in treating x both as an algebraic symbol and as a physical magnitude. Joad asks "What then is it that impinges on the sense organs to start the messages?" He is perplexed because the answer is atoms or things like atoms, which, I have assured him, must not (in exact science) be thought of as possessing any other nature than that of a bundle of pointer-readings. How can a bundle of pointer-readings start a mental process? He might equally ask how can an algebraic symbol x make it difficult to shift an object? The answer is that the inertia or mass which makes the object difficult to move is symbolised by x. And similarly the bundles of pointer-readings symbolise the processes which start the messages. In particular the recurrencies of the pointer-readings stand for recurrencies of the processes.

I pass on to the last of his formal criticisms. "The world of common experience is the datum from which the physicist starts and the criterion by which he judges the validity of the structure he raises. It is therefore presupposed as real and objective throughout."*

* Since the context of this passage refers especially to my discussion of world building in which I stress the effect of mental selection on the characteristics of the physical world, it would be better to substitute "relevance" for "validity", i.e. relevance to the problem of experience.

The argument appears to be that unless a datum is presupposed to be objective no inference can be based on it. This is so astonishing a suggestion that I wonder whether it can possibly be Mr Joad's real opinion. The data furnished by individual experience are clearly subjective, and it is ultimately from these data that the scientific conception of the universe is derived—for what we term "collective experience" is a synthesis of individual experiences. It would seem almost axiomatic that an ultimate datum is necessarily subjective. Joad then goes on to propound as a dilemma (p. 47)—

Thus atoms and quanta are the result of a process of inference based upon observation of the everyday world, while at the same time they originate a process which ends in the construction of the everyday world. Thus the everyday world must be presupposed before the process which results in its construction can take place.

It seems a strange objection to scientific theory that it provides a universe capable of accounting for our everyday experience. Surely the whole intention of inference is that the result of the inference shall be that which is the origin of the datum from which the inference is made. When from an observation of pink rats we infer the presence of alcohol, the validity of the inference lies in the fact that what we infer originates a process which ends in the mental construction of pink rats. Joad's dilemma seems to arise because he gratuitously assumes the presupposition to be "presupposed as objectively real". But it is not presupposed that the pink rats are objectively real.

His difficulty rather suggests that a cyclic scheme of knowledge with which science has familiarised us is not yet appreciated in philosophy. I have formerly* illustrated the nature of a cyclic scheme by a revised version of "The House

* *The Nature of the Physical World*, p. 262.

that Jack Built" which instead of coming to an end repeats itself indefinitely—"...that worried the cat, that killed the rat, that ate the malt, that lay in the house, that was built by the priest all shaven and shorn, that married...". Wherever we start in the cycle we presuppose something that we reach again by following round the cycle. The scheme of physics constitutes such a cycle; and equally we may contemplate a wider cycle embracing that which is beyond physics. Starting at the point of the cycle which corresponds to our individual perceptions, we reach other entities which are constructs from our perceptions. From these we reach other entities, and so on for a number of steps. When we seem to have travelled a long way from our starting point, we find that our perceptions (or more strictly the recurrencies in our perceptions) reappear as constructs from the last-reached entities. The fact that we return by a circuit and not by retracing our steps secures that our adventure is an extension of knowledge and not an excursion in tautology. By the method of Chapter XII we can extract the group structure from the cycle and so express the same truth symbolically without a formal presupposition if we prefer.

IV

The burning question of Determinism is a source of much criticism and controversy. Although the controversial side of the subject is not neglected in Chapter IV, it may be of interest to defend my position with more explicit reference to the views and statements of leading determinists, especially Planck and Einstein. The case against me, based mainly on these authorities, has been ably stated by Sir Herbert Samuel;* the following is taken from an article which I wrote in reply.†

* "Cause, Effect, and Professor Eddington", *The Nineteenth Century and After*, April 1933.
† *Ibid.* June 1933. Reproduced by kind permission of the Editor.

Sir Herbert Samuel has arrested me for trying to rob the public of their most valuable beliefs, and he has placed in the witness-box three of the most eminent physicists now living to give evidence for the prosecution. I suspect that he counts more on the impression that will be produced by this array of authority than on the actual content of their evidence; for there is more protestation than argument in what they have to say. So far as authority is concerned, it would scarcely be possible to name a more formidable trio than Planck, Einstein, Rutherford; nevertheless, I trust that the jury before reaching their verdict will hear patiently what I shall say in my defence.

The occasion of the trial is that I (in common with many modern physicists) have disseminated unbelief in the "Principle of Causality", better known to the public as the doctrine of Determinism. The first designation is generally used by Sir Herbert and his witnesses, but I am not sure that it will be understood by the general reader. I hope the language of the indictment will not lead anyone to suppose that I deny that effects may proceed from causes. The assembly of spectators at an international football match is undoubtedly a cause of the congestion in the streets of Twickenham an hour or so later. But what the principle of causality asserts is that observed causation of this kind is analysable indefinitely, so that each minute movement in the crowd was likewise determined in advance by causes existing hours (or centuries) before. It is this exact and universal causality or predetermination that I challenge.

Of the three witnesses Prof. Max Planck is the one on whom my accuser chiefly relies, and he is the only one whose evidence is in a form which admits of detailed examination. Rutherford, indeed, is too wary to enter into a discussion which might savour of philosophy and takes refuge in a platitude which, though presumably meant for a condemnation of indeterminists, can equally be read as a condemnation of determinists. Planck's views are of special interest because he is the founder of the physical theory which has led to the present crisis; and his arguments are contained in a carefully written book.

In the controversy Determinism *versus* Indeterminism it is essential to have a clear understanding on which side the onus of proof lies—which side is putting forward a *positive* doctrine which it wishes the other side to embrace. Sir Herbert quotes a letter from Lord Rutherford which says:

> It seems to me unscientific and also dangerous to draw far-flung deductions from a theoretical conception which is incapable of experimental verification, either directly or indirectly.

To which side does this apply? My case against Sir Herbert Samuel and his fellow-determinists has been that they develop a far-reaching philosophical outlook based on the principle of causality—a principle which has not been experimentally verified. Here is Einstein's testimony:*

> Hitherto people have looked upon the Principle of Causality as a proposition which would in the course of years admit of experimental proof with an ever-increasing exactitude. Positively defined as a limiting proposition, the principle runs as follows....
> Now Heisenberg has discovered a flaw in the proposition.... The principle of causality loses its significance as an empirical proposition.
> Causality is thus only conceivable as a *Form of the theoretical system*. Now modern physicists are mainly of the opinion that it is inadmissible to build up any sort of theory on what cannot, in principle, be tested.

Einstein, it will be seen, admits that the principle of causality is a positive proposition. He makes no pretence that it has been experimentally verified. Having lost its empirical significance, it is out of range of experimental test

---

* From a letter to Sir Herbert Samuel, published in Sir Herbert's Presidential Address: *Philosophy and the Ordinary Man*, p. 15.

and is indeed only conceivable as a form of *theoretical* system. The words of Lord Rutherford recoil on the prosecution like a boomerang. Out of the mouth of their own witness the principle of causality—the valuable belief of which I am accused of robbing the public—is shown to be "a theoretical conception which is incapable of experimental verification".

Further, compare Planck's testimony:*

> Is it perfectly certain and necessary for human thought that for every event in every instance there must be a corresponding cause?...Of course the answer is in the negative....
> Thus from the outset we can be quite clear about one very important fact, namely, that the validity of the law of causation for the world of reality is a question that cannot be decided on grounds of abstract reasoning.

with Einstein's testimony:†

> Look here. Indeterminism is quite an illogical concept....If I say that the average life-span of such an atom is indetermined in the sense of not being caused, then I am talking nonsense.

Gentlemen of the jury, you have been assured by Planck that it is not a logical necessity of human thought that every event should have a corresponding cause, but nevertheless "physical science, together with astronomy and chemistry and mineralogy, are all based on the strict and universal validity of the principle of causality".‡ Einstein tells you that denial of the principle is illogical, and that it is nonsense to speak of an event as not having a cause; but the principle of causality is a theoretical proposition which, by its very nature, is incapable of experimental test. Rutherford warns you that it is unscientific to base your conclusions on a theoretical conception which is incapable of experimental verification. So now you know just what you are to think

---

* *Where is Science Going?* pp. 112, 113.
† *Ibid.* p. 202.
‡ *Ibid.* p. 147.

of the principle of causality, according to the voice of authority!

Since Planck's discussion is the most extensive, I will treat him as the main witness. I think it is significant of his attitude that he devotes a whole chapter of his book to a survey of the views of different schools of philosophers, whereas the results of physics are accorded less than five pages.* These claim to give the "answer of physics". The crucial paragraph is one already quoted by Sir Herbert Samuel:

> In point of fact, statistical laws are dependent on the assumption of the strict law of causality functioning in each particular case. And the non-fulfilment of the statistical rule in particular cases is not therefore due to the fact that the law of causality is not fulfilled, but rather to the fact that our observations are not sufficiently delicate and accurate to put the law of causality to a direct test in each case. If it were possible to follow the movement of each individual molecule in this very intricate labyrinth of processes, then we should find in each case an exact fulfilment of the dynamical laws.

How does Prof. Planck know this? He speaks as though the whole course of Nature lay revealed to him. Although we cannot apply the test, he knows that the test would be exactly fulfilled if we could apply it. He omits to tell the reader that there is no mention in any modern treatise on quantum theory of the dynamical laws (i.e. causal, as distinct from statistical, laws) to which he here alludes, for the reason that they have not been discovered or even guessed at. Prof. Planck is at liberty to bring this view forward as a hypothesis (if he is prepared to risk Lord Rutherford's displeasure); it is, in fact, the hypothesis usually made by determinists in order to render their doctrine tenable. But to announce it as the answer of science is surely a grave misstatement. Actually the present trend of physical science is against it. I do not

* *Ibid.* pp. 143–147.

mean that it has been disproved; but phenomena which were formerly thought to be a direct consequence of particular causal laws are now acknowledged to be the result of statistical laws, so that they no longer constitute support for Planck's contention. Evidence formerly trumpeted as favourable is now found to be indifferent.

Possibly Prof. Planck intended to stress the first sentence in the above quotation, meaning thereby that it has been proved (mathematically or logically) that unless each individual is governed by strict causal law statistical laws for the assembly are impossible. But does he seriously expect us to believe that the regular experience of life assurance companies would be impossible if the individuals insured had any genuine free will? I think not. I think he is merely stating a practice which used to be followed—of formulating a system of causal law before deducing statistical laws—forgetting that Heisenberg, Schrödinger, Dirac, and others have abandoned this procedure, and that it is their statistical laws which are the basis of existing quantum theory.

Sir Herbert asks whether I am "justified in saying, not that certain scientists, but that *science itself*, has abandoned determinism". I am glad he stresses this distinction. It is illustrated by another of his quotations from Planck:

> Some essential modification seems to be inevitable; but I firmly believe, in company with most physicists, that the quantum hypothesis will eventually find its exact expression in certain equations which will be a more exact formulation of the law of causality.

Thus the causal law is to be found, not in the quantum theory as it is, but in what Planck believes it will eventually become. That is just what I maintain. The law of causality does not exist in science to-day—in that body of systematic knowledge and hypothesis which has been experimentally confirmed. It exists only in the anticipations of certain scientists—anticipations which naturally are coloured by their philosophical predilections.

The philosophical chapter in Planck's book contains one feature which very much concerns our discussion. The chapter begins:*

> This is one of man's oldest riddles. How can the independence of human volition be harmonized with the fact that we are integral parts of a universe which is subject to the rigid order of Nature's laws?
>
> At first sight these two aspects of human existence seem to be logically irreconcilable. On the one hand we have the fact that natural phenomena invariably occur according to the rigid sequence of cause and effect. This is the indispensable postulate of all scientific research.... But on the other hand we have our most direct and intimate source of knowledge, which is the human consciousness, telling us that in the last resort our thought and volition are not subject to this causal order.

The whole chapter is occupied with the various attempts to solve this riddle.

Obviously the riddle does not arise unless we accept the law of causality in Nature. There may be other aspects of the problem of free will leading to other riddles; but the main dilemma, which Planck places in the forefront of the problem, ceases to exist. Many writers have said that our researches into the laws of atomic physics have no bearing on the problem of free will and volition. Planck evidently is not of this way of thinking. For him, as for me, the main problem turns on whether physics does or does not assert the principle of causality.

It is on this point that a number of popular scientific writers have taken up a position that seems to me preposterous. They hold that, since strict causality has not been disproved, and is not incompatible with the new theories, there has not been any modification of the problem. But the dilemma can only

* *Where is Science Going?* p. 107.

be created if physics gives positive support to the principle of causality. It takes two combatants to make a fight—not one combatant and one neutral.

In the present controversy there has been a great tendency to confuse two questions—"Is the law of causality true of the physical universe?" and "Is it the present basis of physical science?" I have quoted Prof. Planck's picture of the mechanism of Nature, which obviously goes far beyond anything warranted by existing knowledge. If I declared this picture to be untrue, I should be open to the same charge of dogmatism as he is. But I can say most assuredly that this picture is not the basis of present-day physics. Present-day physics is simply indifferent to it. We might believe in it to-day and disbelieve in it to-morrow; not a symbol in the modern textbooks of physics would be altered.

Einstein (unlike Planck) fully recognises this change. Whereas Planck holds that modern physics is still based on the law of causality, Einstein recognises that it is not, and he deplores the change. It may be added that under this conviction Einstein has for several years been actively engaged in search for a new theory which shall restore the law of causality to its old supreme position; but hitherto he has not been successful. I need scarcely say that a writer who deals with the philosophical implications of physical science must base his assertions on the existing scheme of knowledge which has resulted from the exertions of Planck, Einstein, Rutherford and others, not on a theory which Einstein hopes some day to produce.

I do not think that the social and political consequences of my teaching will be so terrible as Sir Herbert Samuel fears. He suggests that a student of mine, learning that if he sets light to a barrel of gunpowder an explosion, although highly probable, is not certain, may decide to put the matter to the test. The result will doubtless bear out my assertion that an explosion was exceedingly probable, so I do not see where

CRITICISMS AND CONTROVERSIES 303

the grievance of the relatives of the deceased comes in. And at least I concede to my student freedom to avoid the catastrophe by abstaining from acting in this strange manner; whereas, according to Sir Herbert Samuel and the determinists, the explosion of the barrel is the inevitable outcome of causes which have existed from the beginning of time.

Nor do I think that the substitution of high probability for certainty in the political and economic sphere will be disastrous. It would seem that at the present moment my opportunity for destroying "certainty" in political and economic science is rather limited. Might it not then be better to stress the other side of my conclusions—that, so far as is known, our future is not wholly prearranged by physical law? It is we who have to shape it for better or worse. I have on occasion supported Sir Herbert Samuel and voted for his political efforts for amelioration. My decision was on probability; I could not expect complete certainty that his policy would achieve its end. If any of our leaders can offer the world a solution of the present troubles, we shall not ask for certainty; let him but convince us that the probability of success is—shall we say?—a million to one, and we will follow him to the last ditch.

V

It would be of little advantage to discuss here the controversial aspects of the conclusions which I have reached, or have accepted from the work of others, in regard to purely scientific problems. So far as space permits, I have tried to meet in advance the objections most likely to occur to an attentive reader. But if an expert colleague is unconvinced, or claims to have discovered mistakes and fallacies, the right place to meet him is in a technical journal where mathematical formulae can be countered by mathematical formulae and all our resources for the discovery of truth can be brought to bear. Those who are not prepared to study for themselves the technical arguments, must make what they can of rival assurances that "Codlin's the friend, not Short". I can do no more than pass on such glimpses of illumination as I have found in my own efforts to understand.

There is a point connected with popular expositions of physical science which is perhaps not generally realised. As a rule the results which they translate into non-technical language are obtained partly by strict mathematical deduction, and partly by general arguments as to what hypotheses seem best to accord with physical observation and experiment. Now physics is, or should be, undogmatic; mathematics is, and must be, dogmatic. No mathematician is infallible; he may make mistakes; *but he must not hedge.* Even in this age which dislikes dogma, there is no demand for an undogmatic edition of Euclid; and the examinee who was unable to prove the binomial theorem but "thought he had made it rather plausible" is not held up as an example to be followed. In summarising conclusions for the general reader, mathematical and physical considerations become fused together, and it is impossible to show without elaboration of technical detail where the dogmatic mathematical deduction ends and the plausible physical inference begins. You may therefore find that a book which on the whole reflects the liberal undogmatic attitude of science is chequered with pronouncements which suggest omniscience and intolerance. The latter are a sign (or so it is charitable to assume) that the argument has shifted into the region of strict mathematical deduction, where hedging is not permitted and a definite lead must be given.

Correspondingly there are two kinds of criticism. The one claims to have found a flaw in an author's mathematical deduction; the other dissents from his judgment of the evidence. As to the former, we can only say that one of the parties must be culpably wrong. Supposing, however, that there is agreement on the mathematical side of the problem, there is often room for interesting and valuable controversy on questions of judgment; and some divergence of view is beneficial. Where judgment is more than usually difficult I have tried to indicate the corresponding uncertainty; but there is scarcely any physical conclusion which we can hold as safe from all possibility of revision. Even such a fundamental law as the conservation of energy is now being challenged on account of certain phenomena observed in the production of β rays; I do not myself believe that it is in serious danger, but perhaps I am wrong. On the subject of the constitution of the stars we can scarcely doubt that substantial knowledge has been gained, consideration having been given to all contingencies which we should deem reasonably likely; nevertheless few, if any, of the accepted conclusions, either for the deep interior or for the surface layers of a star, are so unconditional that a star might not evade them if it really wanted to be nasty.

No doubt a detached critic would often recommend suspension of judgment on questions as to which I have ventured to adopt a definite opinion. But I think it would give a wrong picture of scientific activity to view it entirely through such a critic's eyes. The working scientist, like any other man who wishes to accomplish something, must steer a middle course between chronic indecision and precipitant judgment. It is not just a question whether he shall believe this or believe that; it is a choice which may determine whether or not several years of his life shall be spent in working along a blind alley.

VI

My last round will be with Bertrand Russell. I think that he more than any other writer has influenced the development of my philosophical views; and my debt to him is great indeed. But this is necessarily a quarrelsome chapter, and I must protest against the following accusation—*

> Sir Arthur Eddington deduces religion from the fact that atoms do not obey the laws of mathematics. Sir James Jeans deduces it from the fact that they do.

Russell here attributes to me a view of the basis of religion which I have strongly opposed whenever I have touched on the subject. I gather from what precedes this passage that Russell is really referring to my views on free will, which he appears to regard as equivalent to religion; but even so the statement is far from true. I have not suggested that either religion or free will can be deduced from modern physics; I have limited myself to showing that certain difficulties in reconciling them with physics have been removed. If I found a prevailing opinion that Russell could not be a competent mathematician because he had claimed to square the circle, I might, in defending him, point out that the report that he had made such a claim was without foundation. Would it be fair to say that I deduce that Russell is a competent mathematician from the fact that he has not claimed to square the circle?

One might have regarded the foregoing as a casual sacrifice of accuracy to epigram, but other passages make the same kind of accusation:†

> It will be seen that Eddington, in this passage, ‡ does not infer a definite act of creation by a Creator. His only reason for not doing so is that he does not like the idea. The scientific argument leading to the conclusion which he rejects is much stronger than the argument in favour of free will, since that is based on ignorance, whereas the one we are now considering is based upon knowledge. This illustrates the fact that the theological conclusions drawn by scientists from their science are only such as please

---

* *The Scientific Outlook*, p. 112.
† *Ibid.* p. 121.
‡ *The Nature of the Physical World*, p. 83.

them, and not such as their appetite for orthodoxy is insufficient to swallow, although the argument would warrant them.

And again (p. 96):

[Eddington's] optimism is based upon the time-honoured principle that anything which cannot be proved untrue may be assumed to be true, a principle whose falsehood is proved by the fortunes of bookmakers.

Neither my optimism, nor my belief in free will and in religion, nor my belief in Russell's competence as a mathematician is based on this time-honoured principle. But however strong may be the positive grounds for one's opinions, it is not irrelevant to examine the negative grounds and satisfy oneself and others that the evidence which seemed hostile to these beliefs has collapsed.

Memories are short, and one man is sometimes saddled with another man's opinions. It seems worth while therefore to give quotations showing how completely Russell has misstated my view of the relation of science and religion. I think that every book or article in which I have touched on religion is represented in these extracts, except an early essay (1925) which does not provide a passage compact enough to quote.

The starting-point of belief in mystical religion is a conviction of significance or, as I have called it earlier, the sanction of a striving in the consciousness. This must be emphasised because appeal to intuitive conviction of this kind has been the foundation of religion through all ages and I do not wish to give the impression that we have now found something new and more scientific to substitute. I repudiate the idea of proving the distinctive beliefs of religion either from the data of physical science or by the methods of physical science. (*The Nature of the Physical World*, p. 333.)

The lack of finality of scientific theories would be a very serious limitation of our argument, if we had staked much on their permanence. The religious reader may well be content that I have not offered him a God revealed by the quantum theory, and therefore liable to be swept away in the next scientific revolution. (*Ibid.* p. 353.)

It is probably true that the recent changes of scientific thought remove some of the obstacles to a reconciliation of religion with science; but this must be carefully distinguished from any proposal to base religion on scientific discovery. For my own part I am wholly opposed to any such attempt. (*Science and the Unseen World*, p. 45.)

The passages quoted by Mr Cohen make it clear that I do not suggest that the new physics "proves religion" or indeed gives any positive grounds for religious faith. But it gives strong grounds for an idealistic philosophy which, I suggest, is hospitable towards a spiritual religion, it being understood that the guest must provide his own credentials. In short the new conception of the physical universe puts me in a position to defend religion against a particular charge, viz. the charge of being incompatible with physical science. It is not a general panacea against atheism. If this is understood,...it explains my "great readiness to take the present standing of certain theories of physics as being final"; anybody can defend religion against science by speculating on the possibility that science may be mistaken. It explains why I sometimes take the essential truth of religion for granted; the soldier whose task is to defend one side of a fort must assume that the defenders of the other side have not been overwhelmed. (Article in *The Freethinker*.)

I now turn to the question, what must be put into the skeleton scheme of symbols. I have said that physical science stands aloof from this transmutation, and if I say anything positive on this side of the question it is not as a scientist that I claim to speak. (Broadcast Symposium, *Science and Religion*.)

The bearing of physical science on religion is that the scientist has from time to time assumed the duty of signalman and set up warnings of danger—not always unwisely. If I interpret the present situation rightly, a main-line signal which had been standing at danger has now been lowered. But nothing much is going to happen unless there is an engine.

CHAPTER XIV

EPILOGUE

Modern science, in so far as I am familiar with it through my own scientific work, mathematics and physics make the world appear more and more as an open one, as a world not closed but pointing beyond itself.... Science finds itself compelled, at once by the epistemological, the physical and the constructive-mathematical aspect of its own methods and results, to recognise this situation. It remains to be added that science can do no more than show us this open horizon; we must not by including the transcendental sphere attempt to establish anew a closed (though more comprehensive) world.   HERMANN WEYL, *The Open World.*

I

OUR home, the Earth, is the fifth or sixth largest planet belonging to a middle grade star in the Milky Way. Within our galaxy alone there are perhaps a thousand million stars as large and as luminous as the sun; and this galaxy is one of many millions which formed part of the same creation but are now scattering apart. Amid this profusion of worlds there are perhaps other globes that are or have been inhabited by beings as highly developed as Man; but we do not think they are at all common. The present indications seem to be that it is very long odds against a particular star undergoing the kind of accident which gave birth to the solar system. It seems that normally matter collects in big masses with excessively high temperature, and the formation of small cool globes fit for habitation is a rare occurrence. Nature seems to have been intent on a vast evolution of fiery worlds, an epic of milliards of years. As for Man—it seems unfair to be always raking up against Nature her one little inadvertence. By a trifling hitch of machinery—not of any serious consequence in the development of the universe—

some lumps of matter of the wrong size have occasionally been formed. These lack the purifying protection of intense heat or the equally efficacious absolute cold of space. Man is one of the gruesome results of this occasional failure of antiseptic precautions.

To realise the insignificance of our race before the majesty of the universe may be healthful; but it brings to us an alarming thought. For Man is the typical custodian of certain qualities or illusions, which make a vital difference to the significance of things. He displays purpose in an inorganic world of chance. He can represent truth, righteousness, sacrifice. In him there flickers for a few brief years a spark from the divine spirit. Are these of as little account in the universe as he is?

It may be going too far to say that our bodies are pieces of stellar matter which, by a contingency not sufficiently guarded against in Nature, have evaded the normal destiny, and have taken advantage of low temperature conditions to assume unusual complication and perform the series of antics we call "life". I neither assert nor deny this view; but I regard it as so much of an open question that I am unwilling to base my philosophy or my religion on the assumption that it must necessarily break down. But there is another approach to the problem. Science is an attempt to read the cryptogram of experience; it sets in order the facts of sensory experience of human beings. Everyone will agree that this attempt has met with considerable success; but it does not start quite at the beginning of the Problem of Experience. The first question asked about scientific facts and theories, such as we have been discussing in this book, is "Are they true?" I would emphasise that even more significant than the scientific conclusions themselves is the fact that this question so urgently arises about them. The question "Is it true?" changes the complexion of the world of experience —not because it is asked *about* the world, but because it is asked *in* the world. When we go right back to the beginning, the first thing we must recognise in the world of experience is something intent on truth—something to which it matters intensely that beliefs should be true. This is no elusive cryptogram; it is not written in the symbolic language in which we describe the unknowable activities of unknown agents in the physical universe. Before we invite science to take the problem in hand and put in order the facts of sensory experience, we have settled the first ingredient of the world of experience. If science in its survey rediscovers that ingredient, well and good. If not, then science may claim to account for the universe, but what is there to account for science?

What is the ultimate truth about ourselves? Various answers suggest themselves. We are a bit of stellar matter gone wrong. We are physical machinery—puppets that strut and talk and laugh and die as the hand of time pulls the strings beneath. But there is one elementary inescapable answer. *We are that which asks the question.* Whatever else there may be in our nature, responsibility towards truth is one of its attributes. This side of our nature is aloof from the scrutiny of the physicist. I do not think it is sufficiently covered by admitting a mental aspect of our being. It has to do with conscience rather than with consciousness. Concern with truth is one of those things which make up the spiritual nature of Man. There are other constituents of our spiritual nature which are perhaps as self-evident; but it is not so easy to force an admission of their existence. We cannot recognise a problem of experience without at the same time recognising ourselves as truth-seekers involved in the problem. The strange association of soul and body—of responsibility towards truth with a particular group of carbon compounds —is a problem in which we naturally feel intense interest; but it is not an anxious interest, as though the existence of a spiritual significance of experience were hanging in the balance. That significance is to be regarded rather as a datum of the problem; and the solution must fit the data; we must not alter the data to fit an alleged solution.

I do not regard the phenomenon of life (in so far as it can be separated from the phenomenon of consciousness) as necessarily outside the scope of physics and chemistry. Arguments that because a living creature is an organism it *ipso facto* possesses something which can never be understood in terms of physical science do not impress me. I think it is insufficiently recognised that modern theoretical physics is very much concerned with the study of organisation; and from organisation to organism does not seem an impossible stride. But equally it would be foolish to deny the magnitude of the gulf between our understanding of the most complex form of inorganic matter and the simplest form of life. Let us suppose, however, that some day this gulf is bridged, and science is able to show how from the entities of physics creatures might be formed which are counterparts of ourselves even to the point of being endowed with life. The scientist will perhaps point out the nervous mechanism of the creature, its powers of motion, of growth, of reproduction, and end by saying "That's you". But it has yet to satisfy the inescapable test. Is it concerned with truth as I am? Then I will acknowledge that it is indeed myself. The scientist might point to motions in the brain and say that these really mean sensations, emotions, thoughts; and perhaps supply a code to translate the motions into the corresponding thoughts. Even if we could accept this inadequate substitute for consciousness as we intimately know it, we must still protest: "You have shown us a creature which thinks and believes; you have not shown us a creature to whom it *matters* that what it thinks and believes should be true". The inmost ego, possessing what I have called the inescapable attribute, can never be part of the physical world unless we alter the meaning of the word "physical" so as to be synonymous with "spiritual"—a change scarcely to the advantage of clear thinking. But having disowned our supposed double, we can say to the scientist: "If you will hand over this Robot who pretends to be me, and let it be filled with the attribute at present lacking and perhaps other spiritual attributes which I claim as equally self-evident, we may arrive at something that is indeed myself".

A few years ago the suggestion of taking the physically constructed man and adapting him to a spiritual nature by casually adding something, would have been a mere figure of speech—a verbal gliding over of insuperable difficulties. In much the same way we talk loosely of constructing a Robot and then breathing life into it. A Robot is presumably not constructed to bear such last-minute changes of design; it is a delicate piece of mechanism made to work mechanically, and to adapt it to anything else would involve entire reconstruction. To put it crudely, if you want to fill a vessel with anything you must make it hollow, and the old-fashioned material body was not hollow enough to be a receptacle of mental or of spiritual attributes. The result was to place consciousness in the position of an intruder in the physical world. We had to choose between explaining it away as an illusion or perverse misrepresentation of what was really going on in the brain, and admitting an extraneous agent which had power to suspend the regular laws of Nature and asserted itself by brute interference with the atoms and molecules in contact with it.

Our present conception of the physical world is *hollow* enough to hold almost anything. I think the reader will agree. There may indeed be a hint of ribaldry in his hearty assent. What we are dragging to light as the basis of all phenomena is a scheme of symbols connected by mathematical equations. That is what physical reality boils down to when probed by the methods which a physicist can apply. A skeleton scheme of symbols proclaims its own hollowness.

It can be—nay it cries out to be—filled with something that shall transform it from skeleton into substance, from plan into execution, from symbols into an interpretation of the symbols. And if ever the physicist solves the problem of the living body, he should no longer be tempted to point to his result and say "That's you". He should say rather "That is the aggregation of symbols which stands for you in my description and explanation of those of your properties which I can observe and measure. If you claim a deeper insight into your own nature by which you can interpret these symbols—a more intimate knowledge of the reality which I can only deal with by symbolism—you can rest assured that I have no rival interpretation to propose". The skeleton is the contribution of physics to the solution of the Problem of Experience; from the clothing of the skeleton it stands aloof.

II

The scientific conception of the world has come to differ more and more from the commonplace conception until we have been forced to ask ourselves what really is the aim of this scientific transmutation. The doctrine that things are not what they seem is all very well in moderation; but it has proceeded so far that we have to remind ourselves that the world of appearances is the one to which we have actually to adjust our outward lives. That was not always so. At first the progress of scientific thought consisted in correcting gross errors in the familiar conception of things. We learned that the earth was spherical, not flat. That does not refer to some abstract scientific earth, but to the homely earth that we know so well. I do not think any of us have any difficulty in picturing the earth as spherical. I confess that the idea is so familiar to me that it obtrudes itself irrelevantly, and I am liable to visualise a Test Match in Australia as being played upside down. We learned that the earth was rotating. For the most part we give an intellectual assent to this conclusion without attempting to weave it into our familiar conception; but we can picture it if we try. In Rossetti's poem the Blessed Damosel looked down from the golden balcony of Heaven across

> The void, as low as where this earth
> Spins like a fretful midge.

Looking from the abode of truth, perfect truth alone can enter her mind. She must see the earth as it really is—like a whirling insect. But now let us try her with something fairly modern. In Einstein's theory the earth, like other matter, is a curvature of space-time, and what we commonly call the spin of the earth is the ratio of two of the components of curvature. What is the Blessed Damosel to make of that? I am afraid she will have to be a bit of a blue-stocking. Perhaps there is no great harm in that. I am not sure that I would think it derogatory to an angel to accuse him of understanding Einstein's theory. My objection is more serious. If the Blessed Damosel sees the earth in the Einsteinian way she will be seeing truly—I can feel little doubt as to that—but she will be *missing the point*. It is as though we took her to an art gallery, and she (with that painful truthfulness which cannot recognise anything that is not really there) saw ten square yards of yellow paint, five of crimson, and so on.

So long as physics in tinkering with the familiar world was able to retain those aspects which appeal to the aesthetic side of our nature, it might with some show of reason make claim to cover the whole of experience; and those who claimed that there was another, religious aspect of our existence had to fight for their claim. But now that its picture omits so much that is obviously significant, there is no suggestion that it is the whole truth about experience. To make such a claim would bring protest not only from the religiously minded but from all who recognise that Man is not merely a scientific measuring machine.

Physics provides a highly perfected answer to one specialised problem which confronts us in experience. I do not wish to minimise the importance of the problem and the value of the solution. We have seen (p. 11) how in order to focus the problem the various faculties of the observer have been discarded, and even his sensory equipment simplified, until the problem becomes such as our methods are adequate to solve. For the physicist the observer has become a symbol dwelling in a world of symbols. But before ever we handed over the problem to the physicist we had a glimpse of Man as a spirit in an environment akin to his own spirit.

In so far as I refer in these lectures to an experience reaching beyond the symbolic equations of physics I am not drawing on any specialised scientific knowledge; I depend, as anyone might do, on that which is the common inheritance of human thought.

We recognise that the type of knowledge after which physics is striving is much too narrow and specialised to constitute a complete understanding of the environment of the human spirit. A great many aspects of our ordinary life and activity take us outside the outlook of physics. For the most part no controversy arises as to the admissibility and importance of these aspects; we take their validity for granted and adapt our life to them without any deep self-questioning. Any discussion as to whether they are compatible with the truth revealed by physics is purely academic; for whatever the outcome of the discussion, we are not likely to sacrifice them, knowing as we do at the outset that the nature of Man would be incomplete without such outlets. It is therefore somewhat of an anomaly that among the many extra-physical aspects of experience religion alone should be singled out as specially in need of reconciliation with the knowledge contained in science. Why should anyone suppose that all that matters to human nature can be assessed with a measuring rod or expressed in terms of the intersections of world-lines? If defence is needed, the defence of a religious outlook must, I think, take the same form as the defence of an aesthetic outlook. The sanction seems to lie in an inner feeling of growth or achievement found in the exercise of the aesthetic faculty and equally in the exercise of the religious faculty. It is akin to the inner feeling of the scientist which persuades him that through the exercise of another faculty of the mind, namely its reasoning power, we reach something after which the human spirit is bound to strive.

It is by looking into our own nature that we first discover the failure of the physical universe to be co-extensive with our experience of reality. The "something to which truth matters" must surely have a place in reality whatever definition of reality we may adopt. In our own nature, or through the contact of our consciousness with a nature transcending ours, there are other things that claim the same kind of recognition—a sense of beauty, of morality, and finally at the root of all spiritual religion an experience which we describe as the presence of God. In suggesting that these things constitute a spiritual world I am not trying to substantialise them or objectivise them—to make them out other than we find them to be in our experience of them. But I would say that when from the human heart, perplexed with the mystery of existence, the cry goes up, "What is it all about?" it is no true answer to look only at that part of experience which comes to us through certain sensory organs and reply: "It is about atoms and chaos; it is about a universe of fiery globes rolling on to impending doom; it is about tensors and non-commutative algebra". Rather it is about a spirit in which truth has its shrine, with potentialities of self-fulfilment in its response to beauty and right. Shall I not also add that even as light and colour and sound come into our minds at the prompting of a world beyond, so these other stirrings of consciousness come from something which, whether we describe it as beyond or deep within ourselves, is greater than our own personality?

It is the essence of religion that it presents this side of experience as a matter of everyday life. To live in it, we have to grasp it in the form of familiar recognition and not as a series of abstract scientific statements. The man who commonly spoke of his ordinary surroundings in scientific language would be insufferable. If God means anything in our daily lives, I do not think we should feel any disloyalty to truth in speaking and thinking of him unscientifically, any more than in speaking and thinking unscientifically of our human companions.

This attitude may seem to allow too much scope for self-deception. The fear is that when we come to analyse by scientific methods that which we call religious experience, we shall find that the God whom we seem to meet in it is a personification of certain abstract principles. I admit that the application of any method which would ordinarily be called scientific is likely to lead to this result. But what else could we expect? If we confine ourselves to the methods of physical science we shall necessarily obtain the *group-structure* of the religious experience—if it has any. If we follow the less exact sciences they involve the same kind of abstraction and codifying. If our method consists in codifying, what can we possibly obtain but a code? If scientific method is found to reduce God to something like an ethical code, this is a sidelight on the nature of scientific method; I doubt if it throws any light on the nature of God. If the consideration of religious experience in the light of psychology seems to remove from our conception of God every attribute that calls forth worship and devotion, it is well to consider whether something of the same sort has not happened to our human friends after psychology has analysed and scheduled them.

Personification is not necessarily to be condemned as illusory. Am I not myself a personification of that scheme of structure which is all that physical science recognises in me?

III

Let us now consider our answer to the question whether the nature of reality is material or spiritual or a combination of both. I have often indicated my dislike of the word "reality" which so often darkens counsel; but I state the question as it is commonly worded, and answer what I think is in the mind of the querist.

I will first ask another question. Is the ocean composed of water or of waves or of both? Some of my fellow passengers on the Atlantic were emphatically of the opinion that it is composed of waves; but I think the ordinary unprejudiced answer would be that it is composed of water. At least if we declare our belief that the nature of the ocean is aqueous, it is not likely that anyone will challenge us and assert that on the contrary its nature is undulatory, or that it is a dualism part aqueous and part undulatory. Similarly I assert that the nature of all reality is spiritual, not material nor a dualism of matter and spirit. The hypothesis that its nature can be to any degree material does not enter into my reckoning, because as we now understand matter, the putting together of the adjective "material" and the noun "nature" does not make sense.

Interpreting the term material (or more strictly, physical) in the broadest sense as that with which we can become acquainted through sensory experience of the external world, we recognise now that it corresponds to the waves not to the water of the ocean of reality. My answer does not deny the existence of the physical world, any more than the answer that the ocean is made of water denies the existence of ocean waves; only we do not get down to the intrinsic nature of things that way. Like the symbolic world of physics, a wave is a conception which is hollow enough to hold almost anything; we can have waves of water, of air, of aether, and (in quantum theory) waves of probability. So after physics has shown us the waves, we have still to determine the content of the waves by some other avenue of knowledge. If you will understand that the spiritual aspect of experience is to the physical aspect in the same kind of relation as the water to the wave form, I can leave you to draw up your own answer to the question propounded at the beginning of this section and so avoid any verbal misunderstanding. What is more important you will see how easily the two aspects of experience now dovetail together, not contesting each other's place. It is almost as though the modern conception of the physical world had deliberately left room for the reality of spirit and consciousness.

In recognising only two alternatives, material and spiritual, we must naturally employ these terms in a very broad sense. We cannot suppose that the non-material substratum of the physical symbols has elsewhere the specialised development which we recognise in the substratum of the physical symbols which stand for ourselves. But without committing ourselves to any hypothetical generalisation, we can hardly do otherwise than name it spiritual in accordance with the one clue that we have as to its nature.

To see the conception as a whole, consider how you yourself enter into the scheme of knowledge. By scientific investigation I can describe you as part of the physical universe, locate you in space and time, determine your chemical composition, and so on. This is indirect knowledge, for it has come to me (like all my sensory experience) through physical changes propagated along my nervous system. To give this knowledge its most precise form I have to use the symbols of mathematical physics and the equations connecting them. This does not exhaust my knowledge of you. I am convinced that associated with that portion of your brain, which the physiologist identifies more particularly as "you", there is something more. You are not only what these physical symbols describe, but also that "something to which truth matters" whose existence in the world of experience we had to admit from the beginning of our inquiry. I should not be lecturing to you if I were not convinced of this. As an inference, this knowledge of you is even more remote than my knowledge of your physical structure; for it is deduced partly from your physical manifestations and behaviour, and partly from my immediate knowledge of what such manifestations and behaviour imply in my own case. But though the journey is longer, the destination is nearer home. For the knowledge is no longer of the symbolic kind; such a nature as I attribute to you is made up of qualities known to me in my own mind without the intervention of sensory mechanism.

To what extent does this outlook involve the modern conceptions of physics? It is affected in this way. An unreflecting philosophy assumes that the nature of a table is "known to me in my own mind without the intervention of sensory mechanism". Anyone who has the task of expounding the theory of relativity finds himself up against the widespread belief that the nature of space* is known in the mind without the intervention of sensory mechanism. It is due to the relativity theory and the quantum theory that these assumptions have been eradicated from physics, and replaced by the conception of symbolic knowledge which plays so important a part in the argument.

* I do not add Time, because it seems to me that we have *immediate* knowledge of the time sequence in consciousness; and one of the tasks of physics has been to discover the relation between this immediate knowledge of time and our symbolic knowledge of time in the external world obtained through our sensory mechanism. (See *The Nature of the Physical World*, pp. 51, 100.)

It may be asked, Do you then believe that the same spiritual nature which underlies the atoms and electrons in the living brain pervades all atoms and electrons? I would answer that it is inappropriate to speak of atoms and electrons in this connection. We have evidence that your consciousness is associated with a certain portion of your brain; but we do not go on to assume that a particular element of your consciousness is associated with a particular atom in your brain. The elements of consciousness are particular thoughts and feelings; the elements of the brain cell are atoms and electrons; but the two analyses do not run parallel to one another. Whilst therefore I contemplate a spiritual domain underlying the physical world as a whole, I do not think of it as distributed so that to each element of time and space there is a corresponding portion of the spiritual background. My conclusion is that, although for the most part our inquiry into the problem of experience ends in a veil of symbols, there is an immediate knowledge in the minds of conscious beings which lifts the veil in places; and what we discern through these openings is of mental and spiritual nature. Elsewhere we see no more than the veil.

We have travelled far from those comfortable days when, however ignorant we might feel as to the details of the construction of matter, everyone was convinced that he was quite familiar with its essential nature. What are my feelings, my thoughts? What am I myself? Mysteries too deep for the intellect to fathom. What is this table? Oh! Everyone understands that; it is just substance, commonsense reality, reassuringly comprehensible amid the phantasms of our thoughts. No. It is a commonplace reflection that we understand very little about our own minds, but it is here if anywhere that all knowledge begins. As for the external objects, remorselessly dissected by science, they are studied and measured, but they are never *known*. Our pursuit of them has led from solid matter to molecules, from molecules to sparsely scattered electric charges, from electric charges to waves of probability. Whither next?

This does not lead to pure subjectivism. The physical object in the world of my perception is also in the world of your perception. There *is* an external world not part of the mind of either of us, but neutral ground wherein is located the basis of that experience which we hold in common. But I think there can be no doubt that the scientist has a much more mystic conception of the external world than he had in the last century when every scientific "explanation" of phenomena proceeded on the assumption that nothing could be true unless an engineer could make a model of it. The cruder kind of materialism which sought to reduce everything in the universe, inorganic and organic, to a mechanism of fly-wheels or vortices or similar devices has disappeared altogether. Mechanical explanations of gravitation or electricity are laughed at nowadays. You could now safely hand over the human intellect to the tender mercies of the physicist without fear that he would discover in its workings a grinding of cog-wheels. But we must not make too much of these signs of grace in modern physical science. The tyranny of the engineer has been superseded by the tyranny of the mathematician. At least that is a view very widely taken. But alongside this there is a growing realisation that the mathematician is less oppressive a master than the engineer, for he does not claim any insight deeper than his own symbols.

In an earlier book* I have referred to the unconscious habit of the modern physicist of looking on the Creation as though it were the work of a mathematician. Perhaps the irony of these passages is not so evident now as it was at the time. I could not foresee that a few years later a colleague would seriously put forward the view that "from the intrinsic evidence of his creation, the Great Architect of the Universe

* *The Nature of the Physical World*, pp. 104, 209.

now begins to appear as a pure mathematician".* Jeans had previously considered but rejected another explanation. "So, it may be suggested, the mathematician only sees nature through the mathematical blinkers he has fashioned for himself."

In rejecting what seems to me to be the right explanation, Jeans dwells on the failure of anthropomorphic theories and later the devices of the engineer to explain the universe, and he contrasts them with the success of the mathematical conception. There are two factors which, it seems to me, explain the comparative success of the mathematician. In the first place the mathematician is the professional wielder of symbols; he can deal with unknown quantities and even unknown operations. Clearly he is the man to help us to sift a little knowledge from a vast unknown. But the main reason why the mathematician has beaten his rivals is that we have allowed him to dictate the terms of the competition. The fate of every theory of the universe is decided by a numerical test. Does the sum come out right? I am not sure that the mathematician understands this world of ours better than the poet and the mystic. Perhaps it is only that he is better at sums.

IV

The stress here laid on the limitations of physical science will, I hope, not be misunderstood by the reader. There is no suggestion that science has become a declining force; rather we obtain a clearer appreciation of the contribution which it is able to make, both now and in the future, to human development and culture. Within its own limitations physical science has become greatly strengthened by the changes. It has become more sure of its aims—and perhaps less sure of its achievements. Since the last most bewildering revolution of physical theory (wave mechanics) there has been an

* Sir James Jeans, *The Mysterious Universe*, p. 134.

interval of some years during which it has been possible to settle down to steady progress. Recently the most striking developments have been on the experimental side. In quick succession the artificial transmutation of the elements, the discovery of the neutron and the discovery of the positive electron have startled the scientific world and opened up new realms for exploration. But I count this as normal prosperity rather than revolution.

In contemplating the gradually developing scheme of scientific knowledge which never seems to reach finality in any direction, there are times when we are tempted to doubt the substantiality of our gains. Questions, which seem to have been settled, become unsettled:

> Nature and Nature's laws lay hid in night:
> God said, "Let Newton be!" and all was light.
> It did not last. The devil howling, "Ho!
> Let Einstein be!" restored the *status quo*.

In my own subject of astronomy it is particularly difficult to know how far we may feel certain of our ground. So many conclusions have to be guarded by an "if". And it is sometimes those results which have been most widely accepted that prove to have been most insecure. Finding ourselves unable to decide some of those simple fundamental questions, which to a large extent control the course of astronomical theory, we begin to doubt whether there has been any real progress. And then we realise with a start that ten years ago we did not know enough even to formulate the doubts that now beset us. I sometimes think that the progress of knowledge is to be measured not by the questions that it has answered but by the questions that it provokes us to ask.

In writing of the new pathways in science it is natural that the changes should be emphasised rather than the continuity with the past. It may seem that this is an age when we have scant respect for tradition, and are pulling to pieces all that our forerunners so laboriously erected. We have to show unsparingly the way in which the scientists of an earlier generation were misled by false assumptions, and the direction in which their conceptions of the universe have proved inadequate; but we utilise the positive contributions that they made, bringing us step by step nearer to the ideal. Progress has a ruthless side, but it is not an indiscriminate ruthlessness. We are not the less tenderly cherishing the seed planted by our predecessors because from time to time we transplant it into new soil where it may grow more freely. That is what a revolution in science means. When Einstein overthrew Newton's theory, he took Newton's plant, which had outgrown its pot, and transplanted it to a more open field.

All this new growth of science has its roots in the past. If we see farther than our predecessors it is because we stand on their shoulders—and it is not surprising if they receive a few kicks as we scramble up. A new generation is climbing on to the shoulders of the generation to which I belong; and so it will go on. Each phase of the scientific advance has contributed something that is preserved in the succeeding phase. That, indeed, is our ground for hope that the coming generation will find something worth preserving—something that is not wholly illusory—in the scientific thought of the Universe as it stands to-day.

When we see these new developments in perspective they appear as the natural unfolding of a flower:

> For out of olde feldes, as men seith,
> Cometh al this newe corn fro yere to yere;
> And out of olde bokes, in good feith,
> Cometh al this newe science that men lere.

INDEX

A priori probability distribution, 48, 129, 131, 249
Absorption of radiation, 37; in stars, 141; in cosmic cloud, 189, 199
Action, atoms of, 234
Adams, W. S., 155
Aether, 38; mass of, 47
Age, of sun, 165, 167; of universe, 170, 210
Air, 187, 205
Alpha particles, 31, 179
Angular momentum, uncertainty of, 105
Annihilation, of matter, 165, 180; of positrons, 28, 180
Anti-chance, 60, 69
Anticommuting operators, 268, 274
Anti-evolution, 54
Archimedes, 221
Architect of the universe, 323
Aristarchus, 221
Aston, F. W., 182
Atkinson, R. D'E., 179
Atom, structure of, 29, 46, 258; ionisation of, 32, 144
Atomic number, 29

Becoming, 53
Beginning of the world, 58, 60, 67, 220, 306
Bertrand, J., 123
Beta rays, 31, 305
Betelgeuse, 153, 169
Birge, R. T., 251
Blackett, P. M. S., 28
Body and mind, 69, 86, 313

Bohr, N., 72; model atom, 34, 47; correspondence principle, 78
Bond, W. N., 251
Born, M., 72, 82
Bowen, I. S., 204
Brain and mind, 88
Broad, C. D., 75
Broglie, L. de, 41

Calcium, interstellar, 187, 193, 201
Canticles, 84
Causality, law of, 74, 85, 296
Cause and effect, 74, 296
Cavendish experiment, 253
Cavendish laboratory, 144
Cepheid variables, 174, 208
Chance coincidences, 61, 64, 89
Chemistry, law of, 35
Chlorine, isotopes of, 30
Closure of space, 50, 217, 254
Cockcroft, J. D., 160
Coincidences as observational data, 18, 23
Coincidences, chance, 61, 64, 89
Collisions, atomic, 145, 198
Colour sense, 5, 11
Communal objects, 282
Companion of Sirius, 155
Comparison objects and standards, 224
Compressibility of matter, 154, 159
Condorcet, Marquis de, 123
Conservation of energy, 108, 305
Constants of Nature, primitive, 230; numerical, 232
Correspondence principle, 78
Cosmic cloud, 185; density of, 194; temperature of, 198

INDEX

Cosmic rays, 36, 164, 179
Cosmical constant, 47, 214, 220, 222, 230, 247
Cosmical repulsion, 213
Coulomb energy, 236, 240, 243
Creation, 58, 306, 323
Cryptogram, 8, 11, 51
Curvature of space-time, 47, 108, 213, 315
Cyclic scheme of knowledge, 294
Cyclic universe, 59

D line (sodium), 188, 201
Data, initial, 7, 282
Data, sense, 11, 18, 284
Degenerate matter, 158
Dense matter in stars, 155
Detailed balancing, principle of, 57
Determinism, 72, 295; definitions of, 74; and human volition, 86, 301
Deuterium, 31
Deutons, 31
Dingle, H., 20
Dirac, P. A. M., 216, 236, 271, 272, 276, 292
Disorganisation, 55
Dissipation of energy, 66
Distances, table of, 207
Dogmatism, 304
Double star, 187, 241
Dualism, 18
Dwarf and giant stars, 153
Dwarfs, white, 156, 172
Dynamical velocity, 242

E. & O.E., 22
E symbols, 236, 269, 276
Eclipses, prediction of, 83
Ego, 88
Einstein, A., 13, 19, 164, 214, 217, 297, 298, 302
Einstein's law of gravitation, 133, 194

Electromagnetic waves, 36
Electrons, 21, 28; orbits of, 34, 82, 107, 204, 258; wave nature, 42, 101; mass, 243, 247; wave equation of, 227, 247
Elements, number of, 30; transmutation of, 33, 160, 166, 176
Emission of radiation, 37; in nebulae, 203
Empty space, 48
Encounters, atomic, 145, 198
Energy, disorganisation of, 55; source of stellar, 143, 164; conservation of, 108, 305
Energy and mass, equivalence of, 134, 164
Engineer and mathematician, 323
Entropy, 55
Evolution, 54, 58; of elements, 33, 168; of stars, 169
Examination question, 121
Excitation of atoms, 36, 106
Exclusion method, 120, 129, 238
Exclusion principle (Pauli's), 23, 35, 107, 157
Existence, 25, 291
Expansion of space, 215, 218
Expansion of the universe, 67, 211
Experience, problem of, 91, 310
External world, 9, 45, 281, 323

Familiar and scientific stories, 2
Faraday, M., 40
Fermi-Dirac statistics, 239
Field, 39
Field-matter theory, 41, 49
Fine-structure constant, 232, 234, 250
Finite but unbounded space, 50, 108, 217
Fluctuations (of entropy), 63
Fluorescence, 202
Fog (wave mechanics), 42, 246
Forbidden transitions, 204

INDEX

Force, origin of, 213, 240
Forms of existence, 17
Fourth dimension, 275
Fowler, R. H., 156
Fractionating operators, 264
Freedom, degrees of, 242
Free-will, 86, 90, 301, 303
Frequency and probability, 114
Friedman, A., 213, 227
Fürth, R., 253

Galaxies, 206; recession of, 209, 220
Galaxy, rotation of the, 190
Geometrisation of physics, 12
Giant stars, 152, 172
Globe of water, maximum, 194
Gold standard, 81
Gravitation, law of, 133, 194
Group, 262
Group-structure, 256, 274, 318

H line (calcium), 187, 193, 201
Hartmann, J., 187
Heat, nature of, 55, 139; two forms of, 139; maintenance of sun's, 143, 162, 164
Heat-death, 59
Heath, Sir T. L., 221
Heavy hydrogen, 30
Heavy water, 31, 264
Heisenberg, W., 41, 46
Heisenberg's uncertainty principle, 45, 70, 97, 102, 248
Helium nucleus, 31; formation of, 167, 178; isotope of, 178
Henry I, 230
Hertzsprung, 152
Hubble, E. P., 208, 213
Humason, M. L., 212
Hydrogen atom, 30; abundance in stars, 147, 149; transmutation of, 167, 177
Hypersphere, 218, 252

Hypotheses, 20, 266

Idempotency, 263
Ignorable coordinate, 243
Ignorance and probability, 122, 133
Imperfect gas, 154
Impossible and improbable, 64, 79
Indeterminacy of the present, 97, 100, see Uncertainty principle
Indeterminism of the future, 76; amount of, 82, 88, 101
Indeterministic law, 80, 299
Indifference, principle of, 122, 134
Indistinguishable particles, 240, 251
Inert gases, 35
Inference, 9, 92; remote, 5, 280; retrospective, 93; system of, 126
Infinity, 217
Initial data, 7, 282
Instability of atoms, 33, 178; of stars, 174; of universe, 220
Integers, displacement of, 23
Interaction, 238
Interchange, energy of, 240, 243
Interstellar matter, see Cosmic cloud
Interval, 275
Inverse probability, 125, 127
Ionisation, 32, 144; of interstellar matter, 201
Irreducible energy, 108
Irreversible processes, 66
Isotopes, 30; of potassium, 95, 104
Isotropy, origin of, 133

Jabberwocky, 256
Jeans, Sir J. H., 136, 324
Joad, C. E. M., 281, 288
Jumps, orbit, 36, 106, 204, 258

K group of electrons, 34, 37
K line (calcium), 187, 193, 201
Kelvin, Lord, 66, 162

Knight's moves, 259
Knowable universe, 104
Kummer's quartic surface, 271

L group of electrons. 34, 37
Lane, J. H., 135, 138, 152
Laplace, P. S., 74, 110, 195
Larmor, Sir J., 41, 182
Laws of Nature, 8; primary and secondary, 80; causal and statistical, 80, 299
Lemaître, G., 213, 220, 227
Life, 310, 312
Light, *see* Radiation
Limitations of physical science, 315, 324
Lindemann, F. A., 106
Lorentz, H. A., 41
Luminosity of stars, 136, 141, 148, 153

Macroscopic and microscopic theories, 20, 131, 244
Main series, 172
Man, 310, 316
Mass, equivalent to energy, 134, 164; origin of, 108, 249; of electron and proton, 230, 243, 247; of sun, 138; of universe, 221, 248
Mass-defect, 33, 167
Mass-luminosity relation, 153
Mass-ratio, 232, 243, 247, 250
Mathematician and engineer, 323
Maxwell, J. C., 68, 233
Mental realm, 283
Meteors, 164, 202
Metrical field, 39; tensor, 131, 252
Microscopic and macroscopic theories, 20, 131, 244
Mind and time-direction, 52; and entropy, 69; and determinism, 86; and sense-data, 284

Minkowski, H., 275
Models, 266
Momentum, identified with curvature, 47, 108; uncertainty of, 101; angular, 105; of indistinguishable particles, 242
Multillions, 60

Nebulae, gaseous, 184, 196, 202; dark obscuring, 184, 202
Nebulae, spiral, 206; recession of, 209, 220
Nebulium, 203
Negatron, 28, 181
Nerve messages, 3, 7
Neutral realm, 283
Neutron, 31, 32, 181; in stars, 151
New statistics, 157, 239
Nitrogen, 187, 205
Non-technical writing, aim of, 279
Not-hydrogen, 30, 147, 167
Nucleus, atomic, 29; structure of, 32; bombardment of, 33, 176; mass-defect, 33
Number of particles in the universe, 221, 248, 250, 252

Objective reality, 1, 45, 46, 104, 291
Obscuring nebulosity, 184, 202
Observation and theory, 211
Observer, ideal, 13
Occhialini, G. P. S., 28
Oort, J. H., 191
Opacity of stellar matter, 141, 147
Operators (mathematical), 260, 263, 267
Orbits of electrons, 34, 82, 107, 204, 258
Organisation, 55
Organism, 312
Oxygen, 187, 205
Ozone layer, 193